WITHDRAWN

AIR TRAFFIC CONTROL: THE UNCROWDED SKY

GLEN A. GILBERT

AIR TRAFFIC CONTROL:

The Uncrowded Sky

1973   SMITHSONIAN INSTITUTION PRESS
City of Washington

To all Controllers and Pilots "The Air Traffic Control Team" Past, Present, and Future

Copyright © 1973 by the SMITHSONIAN INSTITUTION
Smithsonian Institution Press Publication Number 4873

Library of Congress Cataloging in Publication Data
GILBERT, GLEN ALEXANDER, 1913-
Air traffic control: the uncrowded sky
1. Air traffic control—United States.   I. Title.
TL725.3.T7G523   629.136'6   73-6005
ISBN 0-87474-140-8

Designed by HUBERT LECKIE
Color illustrations by PETE COPELAND
National Air and Space Museum, Smithsonian Institution
Color frontispiece by PETER DE ANNA
National Air and Space Museum, Smithsonian Institution

FOREWORD

Glen Gilbert's specialty in being the outstanding expert and designer of air traffic systems in the Western World makes his treatise on Air Traffic Control all the more important as these systems grow to such crowded proportions.

The widely used winged airplanes of the day—the jet giants—have one very difficult characteristic. They cannot slow down to much less than about two hundred miles an hour when flying near an airport that is congested or when landing or holding preparatory to doing so. This fundamental has ever been in Gilbert's mind and with it goes the importance he assigns to collision avoidance.

The various ways in which this and other problems are attempted to be met by navigating systems devised by Gilbert and many others are detailed in his presentation. In today's concepts, an enormous reliance is placed on radar and radio messages to control towers and centers to enable them to fix the position of planes flying in the air traffic controlled airspace. But it has become evident to many students of this fundamental in our present system that reliance on it under varying weather conditions provides far from the desirable efficiency. There is also a heavy burden placed on the traffic controllers—so much so that their exhaustion is not unusual. Meanwhile, the pilots who have the real responsibility to bring their vessel safely to port, do not actually have the full means to do so.

They are operating, by and large, a vehicle having limited control capability. It cannot be slowed down and stopped in its course. It must charge along at two hundred miles an hour or it ceases to be a safe vehicle. This situation has been accepted by many without realizing that in this element lies one of the greatest barriers that can hamper future serious air transportation growth. Every vehicle from canal boats, bicycles, horse carts, and even sedan chairs can be stopped and backed up in its own element. Boats, trains, autos, hovercraft, are under the control of the captain, conductor, driver, or engineer. Not the airplane—operating in a crowded traffic pattern.

This is not an insurmountable block. We landed on the moon because we had devised a means to land and rise without any forward velocity. Let's do the same with planes in a traffic pattern.

Gilbert has frequently called attention to this concept as being the principal challenging answer to our air traffic problems. *Put more control*—of where he's going—in the hands of the pilot!

GROVER LOENING

CONTENTS

FOREWORD vii
INTRODUCTION xiii
GLOSSARY xv

CHAPTER **I BEHIND THE SCENES** 1

CHAPTER **II HOW IT ALL BEGAN** 8
    Early United States Developments 8
    First Airway Traffic Control Center 9
    Government Operation 9
    Regulatory Actions 10
    Expansion Program 10
    Communication Development 13
    International Impact 13
    Controller Profession 14

CHAPTER **III THE USERS** 15
    Air Carriers 15
    Trunk 16
    Regional 16
    Short Haul 16
    General Aviation 17
    Military 18
    User Needs 19

CHAPTER **IV THE HUMAN ELEMENT** 19
    Controller Preparation 20
    Controller Functions 21
    Controller Pressures 21

CHAPTER **V BASIC PROCEDURES** 23
    Area Control 24
    Flow Control 25
    Approach/Departure Control 28
    Airport Control 28
    Alerting Service 28
    Restrictions 28
    Visual Flight Rules 29
    VFR Separation Standards 29
    Instrument Flight Rules 30
    Separation Standards: General 30
    Separation Standards: Procedural 30

CHAPTER **VI RADAR** 31
    Primary Radar 32
    Basic Principles 32
    Types 32
    What the Controller Sees on Primary Radar Displays 33
    Secondary Surveillance Radar 34
    Basic Principles 34
    Modes 34
    Identification Coding 34
    Altitude Transmission 34
    SSR *System Description* 35
    What the Controller Sees on SSR Displays 35
    Radar Control 35
    Radar Separation Standards 36
    General 36
    Target Separation 36
    SSR Range Accuracy 36
    Radar Separation Minima 36
    General 36
    Passing or Diverging 36
    Adjacent Airspace 37
    Edge of Scope 37
    Obstructions 37
    Final Approach Course Interception 37
    Separation Boxes 37
    Radar Deficiencies 37

CHAPTER **VII AUTOMATION** 38
    Automatic Data Processing 39
    Fail-safe and Fail-soft Concepts 39
    Design Factors 40
    Basic Objectives 40
    Flight Data Processing 40
    Flight Data Updating 41
    Radar Data Processing 41
    Automatic Tracking 41
    Theory of Conflict Prediction and Resolution 41
    Display Generation 42
    Automatic Communications 42
    Basic Features 42
    Air/Ground Communications 43
    Ground/Air Communications 43

CHAPTER **VIII NAVIGATION** 44

Bearing/Distance Navigation 44
VOR/DME Navigation 45
Direction Finding 47
Automatic Direction Finding 47
Hyperbolic Navigation 47
Decca 47
Dectra 48
HARCO 48
LORAN 48
Omega 48
Hybrid 49
Inertial Navigation 49
Doppler Navigation 50
Instrument Landing Systems 50
ILS 50
MLS 51
Basic Instrument Approach Procedures 51
Area Navigation 52
RNAV Benefits 52
Ground-based RNAV 53
Self-contained RNAV 57
Vertical Navigation 57
Time-referenced Navigation 57
Pictorial Displays 58
System-Accuracy Criteria 58
RNAV Flight Plan 60
RNAV Instrument-Approach Procedures 60

CHAPTER **IX AIRCRAFT** 61

Aircraft Categories 61
CTOL Aircraft 62
V/STOL Aircraft 63
Helicopters 63
Folded-Rotor Compound Helicopters 64
Nonhelicopter VTOL 64
Tilt-Wing VTOL 65
Tilt-ducted Propeller VTOL 65
Jet Lift-Engine VTOL 65
Jet Lift-Cruise VTOL 65
STOL Aircraft 65
V/STOL Potential 65
Airport-to-Airport Service 66
Airport-to-City Service 66
City-to-City Service 66
Commuter Service 67
V/STOL Traffic Volume 67
Developing Area Service 68
General Aviation 68
V/STOL Operational Considerations 68
ATC Interface 68
Radar Surveillance 69
Air-derived Separation Assurance 69
Communications 70
All-Weather Operations 70
Personal Aircraft 70
The Supersonic Transport Aircraft 72
Operational Altitudes 72
Range 73
Airports 73
Sonic Boom 73
Airport Noise 73
All-Weather Operations 73
Flight Profile 73
Fuel Considerations 74
SST Maneuverability 74
Special ATC Factors 74
The Hypersonic Transport Aircraft 75

CHAPTER **X AIRPORTS** 75

Airport Function 76
Airport Elements 76
Airport Planning 77
Operational Weather Categories 78
CTOL Airports 78
Runway Capacity 78
Achieving Maximum CTOL Runway Capacity 79
Design Factors 79
ATC *Factors* 80
Lighting 81
Noise 81
Wake Turbulence 82
Heavy Jets 82
Separation Minima 82
Sensing and Suppression 83

Terminal Airspace 83
Basic Terminal Area 83
Terminal Control Area 84
Airport Ground-Traffic Control 84
Surveillance 84
Guidance 84
Subsystem Interaction 84
STOLports 85
Basic Design Criteria 85
STOL Approach/Departure Profiles 86
Other Considerations 86
RTOLports 87
VTOLports 87
Basic Design Criteria 87
VTOL Approach/Departure Profiles 88
Other Considerations 89

CHAPTER **XI COLLISION AVOIDANCE** 91
The Collision Environment 91
Ground-derived Collision Avoidance 92
Air-derived Collision Avoidance 92
Basic Considerations 92
Conspicuity and Vision 93
Pilot Awareness 94
Airborne Detection Instrumentation 94
Pilot Warning Instrument 94
Collision-Avoidance System 95

Air Traffic Situation Display 96
Economic Considerations 96

CHAPTER **XII ONLY THE BEGINNING** 97
Evolution 97
First and Second Generations 97
Third Generation 98
NAS Stage A 98
Automated Radar Terminal System 100
Need for Upgrading 100
Upgraded Third Generation 100
Comparison with Third Generation 101
Analysis 103
Traffic Management 103
Ground Automation 103
Radar 103
DABS 103
Navigation 103
Communications 103
Problem Areas 103
Radar Deficiencies 103
Centralized Traffic Management 104
Pilot Participation 104
Improved Navigation 104
Airport/Airspace Productivity 105
The V/STOL 105
The Next Steps 105
Fourth Generation 106

Controller/Pilot Responsibility 106
Distributed Management Concept 107
Candidate Concepts 107
Supporting Concepts 108
Final Challenge 108

ACKNOWLEDGMENTS 109

INTRODUCTION

Air transportation has become a major element in the overall contribution of transportation to social progress and economic development. It has proven to be one of the most dynamic, partly due to its continuing capability of improving its service and equipment, and partly to the rapid rate at which it is increasing in size and importance in relation to other transportation mediums. It also is apparent that the role of aircraft (air vehicles) as transportation mediums will be influenced by the extent to which they form a part of and are compatible with total transportation system requirements.

Various studies have indicated that a continuing upward trend in aviation progress can be maintained or accelerated only if the numerous facets of air transportation are carefully planned and developed. A number of these facets will be discussed in later chapters of this book, with a view firstly to describe each one as it now functions, including some of its associated problems; secondly, and more importantly, each such facet or element will be discussed in more detail with respect to the manner in which it can contribute to the optimum use of the airspace as a major transportation medium.

While several of these elements which comprise air transportation have been identified by some as being factors which may limit the continued advancement of the air transportation system, it should be recognized that this is not necessarily the case. For example, "the airport bottleneck" is frequently blamed as an excuse for extensive air traffic delays and as an incurable factor limiting air service capacity. However, there are thousands of airports all over the world which are seldom used, and the productivity of most existing airports can be increased by applying various corrective techniques. Lack of airspace—the "Crowded Sky" syndrome—also is sometimes called a factor limiting the continued development of air transportation. But the fact is that we have not begun to use the potential of the airspace as a transportation medium! It is the system that is crowded—not the sky! In other words, our objective must be to learn how to utilize most effectively the virtually unlimited capacity of our **Uncrowded Sky.**

What really is needed in order to foster the expansion of the air transportation system to meet public demand in a total transportation concept is to apply more effective methods to achieve the highest possible degree of efficiency in the use of airports and airspace. This objective depends to a large extent on the capability of the overall Air Traffic Control (ATC) System. Measures to achieve it are discussed throughout this book.

The objectives of an air traffic control (air traffic management) service, as defined by the International Civil Aviation Organization (ICAO), are to:

Prevent collisions between aircraft in flight

Prevent collisions between aircraft on the maneuvering area of an airport and obstructions on that area

Expedite and maintain an orderly flow of air traffic

Provide advice and information useful for the safe and efficient conduct of flights

Notify appropriate organizations regarding aircraft in need of search and rescue aid, and assist such organizations as required.

The basic facilities employed to perform the air traffic control service are:

Centers which provide en route and area traffic control service

Towers which control traffic on and in the vicinity of airports

Flight service stations which feed flight plan information into centers and towers as well as providing briefing services to pilots.

The degree of air traffic control exercised by these facilities depends basically on the meteorological conditions in which an aircraft is flown. Broadly speaking, when an aircraft can be flown clear of clouds and the pilot has good visibility, the flight is conducted in accordance with "visual flight rules" and is referred to as a "VFR" flight. VFR flights are subject to little or no control by the ground facilities. It is up to the pilot who has been qualified for only VFR flight to watch out for the safety of his flight in the "see and be seen" environment. If a flight cannot be conducted in accordance with VFR, it must be conducted under "instrument flight rules"—"IFR," and the ground facilities exercise positive separation control over all such flights. Even though a flight can be conducted in VFR conditions, the pilot may request IFR control (if he is qualified for instrument flight and his aircraft has the proper airborne equipment) from the ground facilities should he desire to receive full Air Traffic Control Service.

In the overall concept, the ATC System includes:
Aircraft
Airports
Navigation and communication services
Airborne and ground equipment
Weather services
Rules, regulations and procedures
Controllers and pilots.

Since its inception in the mid-1930s, Air Traffic Control has been a dynamic service with an ever-growing series of problems. As each new type of aircraft became integrated into the world's air transportation system, new ATC techniques, procedures, and equipment have been required. Thus, ATC systems have been designed to a large extent around the typical aircraft and air traffic volume at various periods in time.

During the period 1935-1955, for example, the cruising speeds of the aircraft in use were, in general, not much more than 200 knots, and traffic volume was relatively low. These factors provided an environment in which the Air Traffic Control System could function somewhat leisurely, relatively speaking, with minimum traffic congestion problems and on a "manually" operated basis.

With the passing years, however, the ATC environment changed very radically. The number of aircraft in the system increased at a high rate with a constant upward trend. Similarly, aircraft speeds continued to increase in some categories like the airline and business jets; whereas other types, such as some private aircraft, remained in the much lower speed ranges. Yet, the ATC System needed to be so designed as to permit *all* categories of airspace users to operate in the airspace and to and from all airports. New equipment was increasingly required to meet changing demands, such as radar, improved navigation and communication facilities, and automation of manually performed functions. New air vehicles began to emerge from the drawing boards—from the vertical/short takeoff and landing aircraft (V/STOLs) to the supersonic transports (SST's).

Thus, an evolutionary development concept for improving the ATC System has resulted as a natural consequence of the changing environment within which the ATC System must be capable of functioning. This evolutionary concept in turn has led to a series of ATC System generations. These have commenced approximately every ten years or so, with the first generation beginning about 1935 in the United States. Each generation lasts about twenty years, having an overlap between one another for a number of years (± 10) during the transition periods.

For example, the United States' second generation ATC System started about 1950, and the first generation was phased out in the mid to late 1950s. The third generation commenced in the early 1960s, and an upgraded third generation was initiated in the early 1970s. The latter two generations will merge during the mid to late 1970s and probably continue into the 1990s. A fourth generation (also referred to as an Advanced Air Traffic Management System) is scheduled to get underway in the mid-1980s and to last into the next century.

The evolutionary development of the ATC System, in general, follows similar lines in many other parts of the world. Although this book deals essentially with the Air Traffic Control System of the United States, the basic principles and problems apply on a worldwide basis. The fundamental challenge to the ATC System is the same everywhere—to provide the means which will make possible the safe and expeditious use of the airspace by all who desire to use it as a transportation medium.

GLOSSARY

ADEU	automatic data-entry unit	CW	continuous wave	HST	hypersonic transport
ADF	automatic direction finding	DABS	discrete address beacon system	IAF	initial approach fix
AGL	above ground level	Decca	a British hyperbolic navigation system similar in geometrical principle to LORAN	IAP	instrument approach procedure
AGTC	airport ground-traffic control			ICAO	International Civil Aviation Organization
AMST	advanced medium STOL transport	Dectra	a hyperbolic navigation system for use over large bodies of water	ICAN	International Convention for Air Navigation
ARINC	Aeronautical Radio, Inc.			ICNI	integrated communication, navigation, and identification
ARSR	air route surveillance radar	DF	direction finding		
ARTS	automated radar terminal system	DH	decision height	IFATCA	International Federation of Air Traffic Controllers Associations
ASDE	airport surface detection equipment	DME	distance measuring equipment		
ASR	airport surveillance radar	DOC	direct operating costs	IFR	instrument flight rules
ATC	air traffic control	DVOR	Doppler very-high-frequency omnidirectional range (VOR)	ILS	instrument landing system
ATCAC	Air Traffic Control Advisory Committee			INS	inertial navigation system
ATCRBS	ATC radar beacon system (also referred to as secondary surveillance radar—SSR)	ELT	emergency locator transmitter	IOCE	input-output control element
		EPNdB	effective perceived noise in decibels	IPS	intermittent positive control
ATIS	automated terminal information service	Eurocontrol	a multi-national organization controlling air traffic in Western Europe	ISA	international standard atmospheric conditions
ATSD	air traffic situation display				
bit	BINARY DIGIT	FAA	Federal Aviation Agency/Administration	LF	low frequency
CAR	Civil Air Regulations	FAF	final approach fix	LORAN	LONG RANGE NAVIGATION
CAA	Civil Aeronautics Authority	FAR	Federal Air Regulations	MAWP	missed-approach waypoint
CARF	central altitude reservation facility	FDP	flight data processing	MDA	minimum descent altitude
CAS	collision-avoidance system	FDSU	flight data storage unit	MLS	microwave landing system
CCC	central computor complex	FLAT	flight-plan-aid tracking	MSL	mean sea level
CDI	course deviation indicator	FSS	flight service station	MTI	moving target indicator
CE	computing element	GCA	ground controlled approach	NAS Stage A	national airspace system—"Stage A" automation
CFR	contact flight rules	HAA	height above airport		
CP	circular polarization	HARCO	HYPERBOLIC AREA COVERAGE system	NDB	nondirectional beacon
CRT	cathode ray tube	HAT	height above touchdown	NOTAM	notice to airmen
CTOL	conventional takeoff and landing	HSI	horizontal situation indicator	OCA	oceanic control areas

PAM	peripheral adapter module	STOLAND	STOL navigation and landing system
PAR	precision-approach radar	STOLport	a specially designed airport used by short takeoff and landing aircraft
PCA	positive control area		
PDME	precision distance measuring equipment	TACAN	TActical AIr Navigation
PICAO	Provisional International Civil Aviation Organization	TCA	terminal control area
		TCU	tape control unit
PPI	plan-position indicator	TDZ	touchdown zone
PVOR	precision very-high-frequency omnidirectional range (VOR)	T/F	time frequency
		TRACON	terminal control (facility)
PWI	proximity warning instrument	TVOR	terminal (area) very-high-frequency omnidirection range (VOR)
radar	RAdio Detection And Ranging		
RCAG	remote communications air-ground (facility)	UHF	ultra high frequency
		VASI	visual-approach slope indicator
RDP	radar-data processing	VTOLport	a specially designed airport used by vertical takeoff and landing aircraft
REIL	runway-end identifier lights		
RNAV	area navigation	VFR	visual flight rules
RSS	root-sum-square	VHF	very high frequency
RTOL	reduced takeoff and landing	VLF	very low frequency
RVR	runway visual range	VNAV	vertical navigation
SE	storage element	VOR	very-high-frequency omnidirectional range
SID	standard instrument departure		
special VFR	special visual flight rules	VOR/DME	very-high-frequency omnidirectional range/distance measuring equipment
SPI	symbolic pictorial indicator		
scramjet	supersonic combustion ramjet	VORTAC	ground radio navigation station combining VOR/DME and TACAN
SSR	secondary surveillance radar		
SST	supersonic transport	VSI	velocity and steering indicator
STAR	standard terminal arrival route	V/STOL	vertical and/or short takeoff and landing
STOL	short takeoff and landing	VTOL	vertical takeoff and landing

1 Dulles Tower structure, includes local control functions in the "cab" in the top of the structure, with departure control, ground control, approach control, and radar surveillance in the several floors beneath.

CHAPTER I

BEHIND THE SCENES

"Ladies and Gentlemen, this is your Captain speaking. As you make yourselves comfortable on United National Airlines Flight 9, from Dulles International Airport, Washington, D.C., to Los Angeles International Airport at Los Angeles, California, you are going to be 'invisible members' of our crew and sit with us on the flight deck. As a special feature of this VIP flight, courtesy of the Smithsonian Institution, one of our reserve captains is going to be your observer on the flight deck. He will keep you informed as to what is going on 'behind the scenes' with the organization on the ground operated by the Federal Aviation Administration—Air Traffic Control, or "ATC"—which will safeguard us from any possible collision with other aircraft and expedite our flight from departure to destination.

We are about to leave the loading ramp at Dulles. Our flight time to Los Angeles is estimated to be five hours and twenty-five minutes. From now on the flight crew will be in direct radio communication continuously with the ground air traffic controllers. Communication links with them are maintained via remotely located radio communication air/ground facilities tied into the control towers at departure and destination airports as well as with the control centers while enroute. In addition, we can communicate directly at any time with FAA communications personnel at flight service stations associated with the ground navigation facilities we will be using, and with our own company's dispatchers using an aeronautical radio system owned and operated by the airlines. So, just stay tuned in and watch what goes on. Listen to our radio communications with the ground controllers along with explanatory comments by our guest cockpit observer. The First Officer is about to call Dulles Tower for taxiing instructions."

Pilot: Dulles Clearance Delivery. This is United National 9, IFR to Los Angeles. Over.

Controller: United National 9. This is Dulles Clearance Delivery. Cleared as filed. Maintain four thousand, expect flight level two three zero passing Cassanova. Squawk one zero two niner. Contact Dulles Departure Control on one two five point five. Fly runway heading after airborne for vectors to Cassanova. Over.

Observer: The Captain has been authorized by the ATC clearance delivery controller to proceed as per previously filed flight plan under fully controlled (IFR) air traffic procedures. After takeoff, we will climb to 4,000 feet and hold this altitude until reaching a ground navigation fix called "Cassanova" which is about 25 miles southwest of Dulles. Then we may expect to climb to a flight level of 23,000 feet. During the flight our aircraft will be identified by the number 1029 derived from our airborne radar beacon. During our trip we will be asked to "ident" which means that the pilot presses a button to activate the aircraft's radar beacon identification. This, then, makes a distinctive display on the ground

2 Inside the Dulles Tower.

controller's radarscope, thus positively identifying our flight.

We are now about to switch communication frequency to 125.5 mHz to contact the Dulles departure controller, and have been instructed after takeoff to maintain the takeoff runway heading until we receive new headings or radar 'vectors' and arrive at the Cassanova fix. The pilot is about to acknowledge the instructions for verification by the clearance delivery controller, after which he will be instructed to shift to the communication frequency of 121.9 mHz to receive taxiing instructions from the ground controller.

Pilot: United National 9 is cleared as filed. Maintain four thousand, expect flight level two three zero passing Cassanova. Squawk one zero two niner. Contact Dulles Departure Control on one two five point five. Fly runway heading after airborne for vectors to Cassanova.

Controller: United National 9. Clearance correct. Contact ground control on one two one point niner when ready to taxi.

Observer: At this point, the pilot is tuning his radio to the automated terminal information service called "ATIS," which is a continuous broadcast of recorded noncontrol information. These broadcasts include specific traffic pattern information, type of instrument approach to be expected (for arriving aircraft,) runways to use, surface winds, ceiling and visibility, the altimeter setting, and the temperature, and are currently identified by a letter in a phonetic code, such as "Bravo" for "B." We are now tuned into the ATIS frequency: *"This is Dulles International Airport Information Bravo. Dulles weather ceiling measured eight hundred overcast, visibility four, light rain and fog, temperature three niner, wind zero two zero degrees at one zero, altimeter two niner seven niner, ILS. One right approaches are in progress, departure runways one left and runway three zero. Advise the arrival or ground controller on initial contact that you have received information Bravo."*

After receiving the ATIS information, the pilot now tunes his radio to the ground control frequency and is ready to taxi. Let's listen in.

Pilot: Dulles Ground Control. This is United National 9 at ramp two zero, ready to taxi, IFR flight to Los Angeles. Over.

Controller: United National 9. This is Dulles Ground Control. Taxi to runway three zero. Proceed westbound on R two, turn left on W one. Over.

Pilot: United National 9. Roger.

Observer: Your captain is now taxiing toward the departure end of the runway 30 (heading 300°) under the watchful eye of the ground controller. As our flight arrives at the departure end of the runway, the pilot is tuning his transmitter to the local controller in the "cab" on top of the tower. He is now about to request clearance for departure.

Pilot: Dulles Tower. This is United National 9 ready for takeoff. Runway three zero. Over.

Controller: United National 9. Roger. Hold short of runway three zero. Traffic a DC-9 on short final. Over.

Pilot: United National 9 holding short.

Observer: The DC-9 which was on a short final has now landed. The controller is instructing us to taxi into position for takeoff on the runway and it goes like this.

Controller: United National 9 taxi into position and hold. Runway three zero. Over.

Pilot: United National 9 cleared into position, Runway three zero. Roger.

Observer: The DC-9 is now completely clear of the runway and we're ready to go.

Controller: United National 9 cleared for takeoff. Maintain runway heading for vectors to Cassanova. Over.

Pilot: United National 9. Roger. We're rolling.

Observer: As you see we are now airborne, and the local controller in the tower is about to pass us on to the departure controller.

Controller: United National 9 contact departure control now. Over.

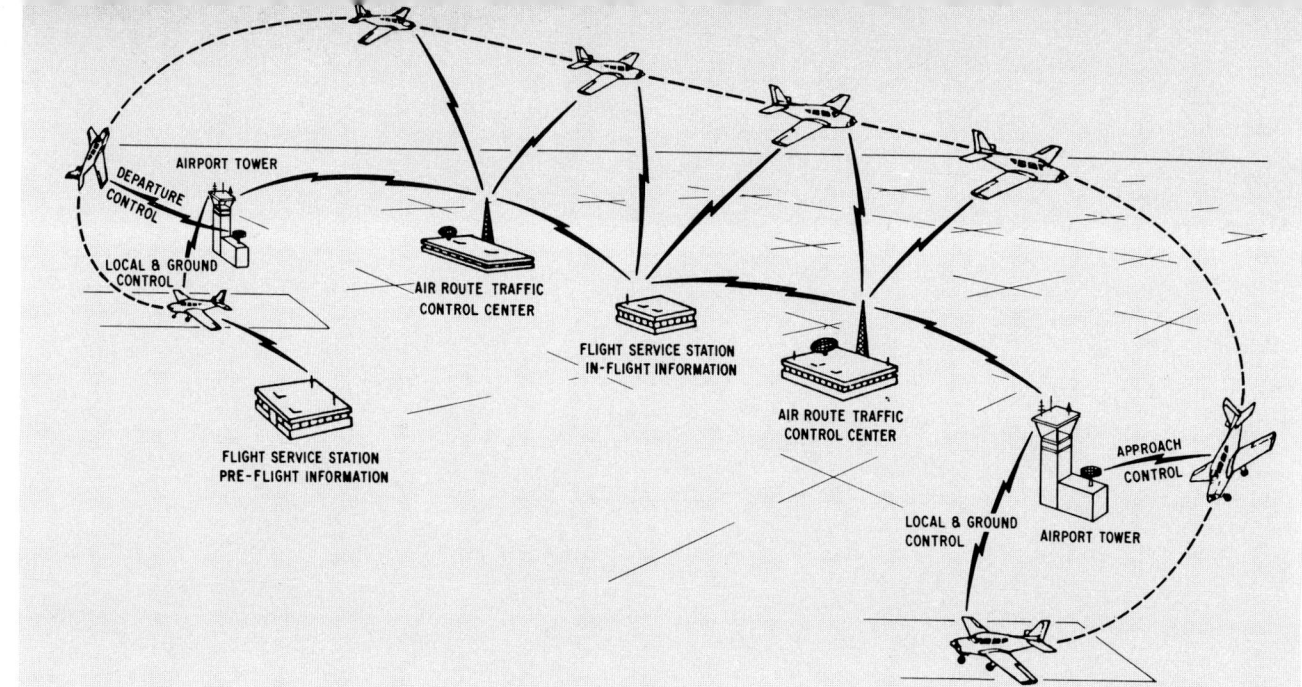

3 Functions of the Air Traffic Control System are carried out by specialized ground-control facilities interconnected with each other and in constant communication with the controlled aircraft.

Pilot: United National 9. Roger. Good-day.

Pilot: Dulles Departure Control. This is United National 9 with you; leaving one thousand, climbing to four thousand. Over.

Observer: We are proceeding as per our previously approved clearance and will shortly be receiving headings to follow in degrees — called "vectors." These instructions are based on radar displays which the controllers are watching constantly to be sure that our aircraft has safe separation from other air traffic. These vectors will lead us to our enroute flight path which is shown on aeronautical charts and is identified by the letter "J" (which stands for "jet route") followed by the identification number of the route.

Controller: United National 9. This is Dulles Departure Control. Radar contact. Turn left, heading two four zero for vectors to intercept Jay one forty nine. Over.

Pilot: United National 9. Roger. Turn left, heading two four zero.

Observer: At this point, let's project ourselves into the ground ATC System for a moment to see what the controllers involved with our flight are doing. The Dulles departure controller is calling the Washington Air Route Traffic Control Center. The purpose of this call is to initiate the radar "handoff" by which radar identification of an aircraft is made known from one controller to another without interruption of radar flight following. Radar handoffs will take place approximately twenty-five times as our flight traverses one controller's area of responsibility to that of another. The following is typical of the dialogue that would take place between controllers during a radar handoff using direct telephone communication channels.

Dulles Departure Controller: Washington Center. Dulles Departure with a handoff.

Washington Center Controller: Washington Center, go ahead.

Dulles Departure Controller: United National 9 is ten miles east northeast of Cassanova level at four thousand.

Washington Center Controller: Radar contact United National 9, flight level three five zero is approved. Bravo Alpha.

Dulles Departure Controller: Oscar Juliet.

Observer: In inter-ATC communications, words such as Bravo Alpha and Oscar Juliet are called controller operating initials. These are stated by controllers at the end of completed radar handoffs and have a certain significance such as transmission completed and acknowledgment of receipt of another controller's transmission. With the inter-controller coordination now completed, responsibility for our flight is about to be transferred or "handed off" to the Washington Center.

Controller: Dulles to United National 9. Climb and maintain flight level three five zero, contact Washington Center one three four point three. Over.

Pilot: United National 9. Roger. Climbing from four thousand to flight level three five zero. Contacting Washington Center one three four point three. Good-day.

Pilot: Washington Center. This is United National 9. Leaving five thousand, climbing to flight level three five zero. Over.

Controller: United National 9. This is Washington Center. Roger. Ident. Continue present heading until intercepting Jay one forty nine, then resume normal navigation. Report leaving flight level two one zero. Over.

Pilot: United National 9. Roger. Report leaving flight level two one zero.

Observer: We have been cleared to our cruising level of 35,000 feet, and the pilot has just advised the Washington Center that we are leaving 5,000 feet climbing to 35,000 feet. The Washington Center has now taken over radar surveillance of our aircraft, and we are continuing our last radar vector until we intercept the jet airway to Los Angeles. Once we are on this airway, the pilot will navigate the aircraft by reference to cockpit instruments based on inputs from ground radio navigation stations called "VORTACS." These stations transmit multiple radials which are like spokes on a wheel.

4 Typical air route traffic control center.

We will follow a selected radial to and from a number of VORTACs during our flight to Los Angeles. Using distance measuring equipment, the pilot knows at all times how far he is from a given ground station. The Washington Center is now about to transfer our flight from one sector of its control area to another by changing communication frequency. Here is the way the intra-center handoff goes.

Pilot: Washington Center. United National 9 leaving flight level two one zero.

Controller: United National 9. Roger. Contact Washington Center one two seven point six now. Over.

Pilot: Roger. One two seven point six.

Pilot: Washington Center. United National 9 leaving flight level two two zero. Over.

Controller: United National 9. This is Washington Center. Roger. Ident. Verify assigned altitude as flight level three five zero. Over.

Pilot: Roger. United National 9 is climbing to flight level three five zero and we'll call you when we get there. Over.

Controller: United National 9. Roger.

Pilot: Washington Center. United National 9 level at flight level three five zero. Over.

Controller: United National 9. Roger.

Observer: We have now reached our cruising altitude of 35,000 feet and are on our assigned airway. The various instructions which we have been receiving from the ground controllers have in effect provided a special flight path through the airspace for our aircraft to keep us clear of other air traffic. As we proceed enroute, should you happen to notice any other aircraft you will know that a ground controller is looking out for us. Now, we are about to leave the Washington Center's control area and are about to be handed off to the Indianapolis Center.

Controller: United National 9, contact Indianapolis Center one three four point seven five now. Over.

Pilot: United National 9. Roger. One three four point seven five. Good-day.

Pilot: Indianapolis Center. United National 9. Flight level three five zero.

Controller: United National 9. This is Indianapolis Center. Roger. Ident.

Pilot: Indianapolis Center United National 9. We are experiencing light to moderate rough air. We would like to get a higher altitude to give the passengers a smoother ride. Over.

Controller: United National 9. Roger. Climb and maintain flight level three niner zero and report when reaching. Over.

Pilot: United National 9. Roger. Leaving flight level three five zero for flight level three niner zero.

Pilot: United National 9 is level at flight level three niner zero.

Controller: United National 9. Roger.

Observer: You will note that when we encountered some turbulence at our previously approved cruising altitude, the Captain requested and received from the controller permission to climb to 39,000 feet. In other words, the controller determined by reference to his radar display that there was no other air traffic which would conflict with our flight during climb to the new altitude. As we leave the jurisdiction of the Indianapolis Center, we will be handed off to the next center.

Controller: United National 9, contact Kansas City Center on one three four point three. Over.

Pilot: United National 9. Roger. One three four point three.

Pilot: Kansas City Center. This is United National 9. Flight level three niner zero. Over.

Controller: United National 9. This is Kansas City Center. Roger. Ident. Over.

Observer: The Captain has just noted some bad weather ahead and is about to request a change of altitude from the Kansas City Center.

Pilot: Kansas City Center. This is United National 9. We see some weather ahead on our radar and we would like to try a little lower altitude, if possible. Over.

Controller: United National 9. Roger. Descend and maintain flight level three one zero. Over.

Pilot: United National 9. Roger. Out of flight level three niner zero for three one zero.

Pilot: United National 9. Level at flight level three one zero.

Observer: Well, we are now moving into the next to last enroute-center's control area on our flight to Los Angeles, that of the Denver Center. Let's listen to what the Denver Center has to say.

Controller: United National 9. Roger. Contact Denver Center one three five point one five. Over.

Pilot: Roger. Contact Denver Center one three five point one five. Good-day.

Pilot: Denver Center. This is United National 9. Level flight at flight level three one zero.

Controller: United National 9. This is Denver Center. Roger. Ident. Over.

Controller: United National 9. We have just been advised that the present weather in Los Angeles is ceiling indefinite obscured. Visibility one-eighth. Delays and congestion in the area do not permit your flight to continue as scheduled. Your clearance limit is now Peach Springs VORTAC. I say again, Peach Springs VORTAC. Maintain flight level three one zero. Expect further clearance at two one zero eight. Over.

Pilot: Roger. Cleared to Peach Springs VORTAC. Expect further clearance two one zero eight.

Observer: We are continuing to fly at our last assigned altitude — 31,000 feet, but cannot pass the Peach Springs VORTAC station, which is about 240 miles east northeast of Los Angeles, until further cleared by ATC. We may expect such further clearance at 2108 Greenwich time. This means we will probably have about a twenty-minute delay in our arrival at Los Angeles.

Observer (later): You may have noticed that we have been following an eliptical flight path for the past twenty minutes or so. This is known as "holding." This procedure is specified by ATC in conditions of traffic congestion in order to set up proper spacing between the different aircraft for approach into the terminal area. We are now receiving a call from the Denver Center.

Controller: United National 9. Contact Los Angeles Center one three four point niner five. Over.

Pilot: Roger. Contact Los Angeles Center one three four point niner five. Good-day.

Pilot: Los Angeles Center. This is United National 9. Flight level three one zero. Over.

Controller: United National 9. This is Los Angeles Center. Roger. Ident. You are now cleared to the Los Angeles Airport over Peach Springs VORTAC, via the last routing cleared. Maintain flight level three one zero. Over.

Pilot: Roger. Cleared to Los Angeles Airport via the Peach Springs VORTAC. Then via the last assigned routing. Maintain flight level three one zero.

Observer: We are now beginning to approach the Los Angeles terminal area. Due to the weather and a heavy volume of traffic, we may expect to be vectored and given "speed controls." These procedures are applied by the ground controllers to achieve not only safe separation from other aircraft, but also to expedite traffic flow and minimize delay by metering the various aircraft at optimum spacing for landing at the airport. We will follow this process involving several handoffs between sectors of the Los Angeles Center and with different control positions at the Los Angeles Tower.

Controller: United National 9. Descend now and maintain one zero thousand. Altimeter setting is two niner seven six. Over.

Pilot: Roger. United National 9 is cleared to ten thousand. Altimeter two niner seven six.

Controller: United National 9, contact Los Angeles Center one two eight point two now. Over.

Pilot: United National 9. Roger. One two eight point two.

Pilot: Los Angeles Center. This is United National 9.

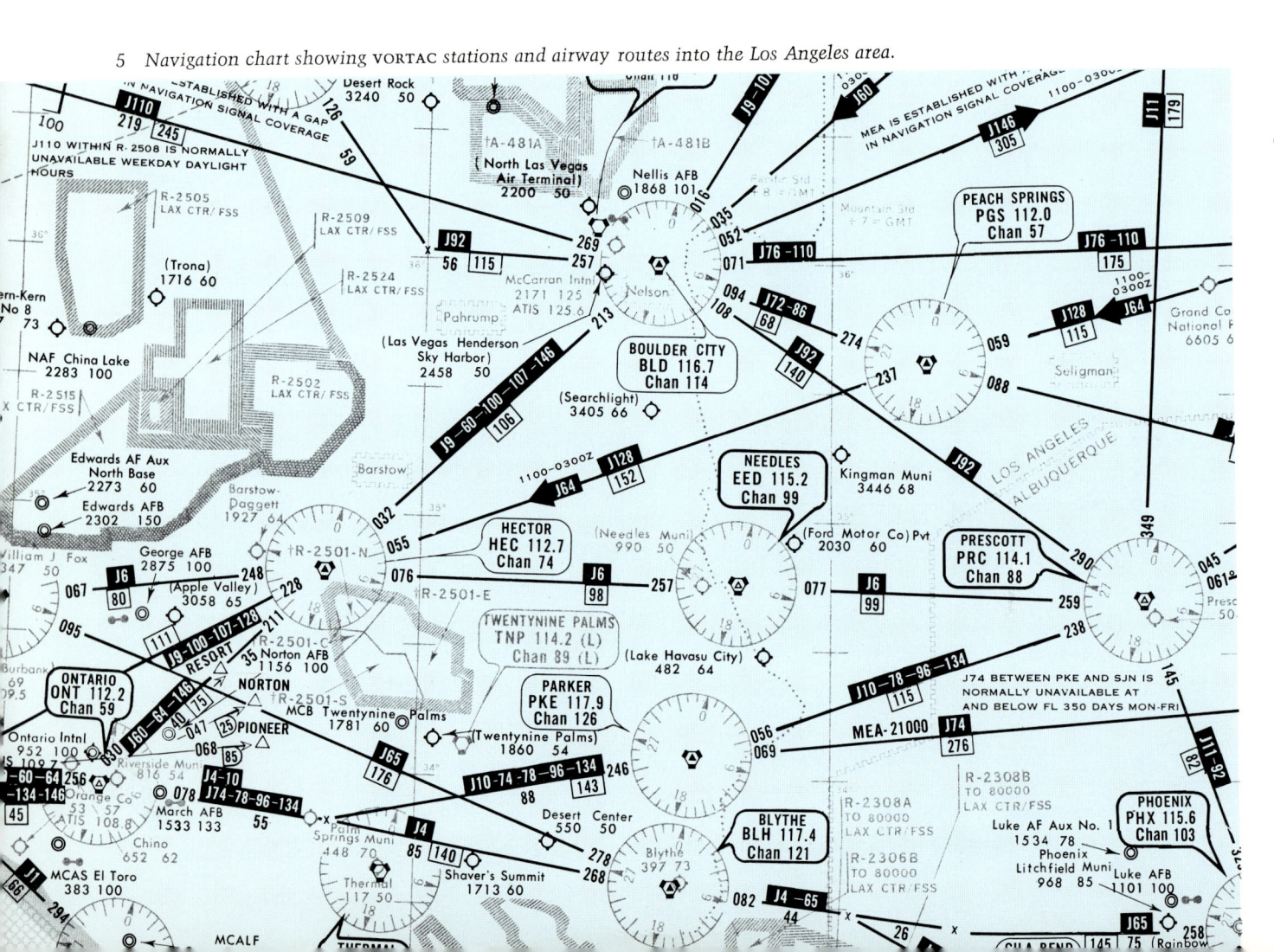

5 Navigation chart showing VORTAC stations and airway routes into the Los Angeles area.

Leaving flight level two two zero for one zero thousand. Over.

Controller: United National 9. This is Los Angeles Center. Roger. Ident. Report leaving one three thousand. Over.

Pilot: United National 9. Roger. Call leaving one three thousand.

Pilot: Los Angeles Center. United National 9 passing one three thousand.

Controller: United National 9. Roger. Squawk zero four zero zero and contact Los Angeles Approach Control one two four point niner now. Over.

Pilot: United National 9. Roger. Squawking zero four zero zero. Good-day.

Observer: We are about to enter the jurisdiction of the approach control facility associated with the Los Angeles Tower. Here, the controllers will be using the latest automated radar terminal system called "ARTS-III" which shows them continuously our flight identification, position, altitude, and ground speed. Their control area extends over a radius of about forty miles from the Los Angeles Airport. As we enter the terminal area, we have tuned in Los Angeles ATIS and have listened to information Zulu. This gave us the latest weather, runway in use, altimeter setting, and other nontraffic control information. We now have been

6 Northwesterly view across 3,500-acre Los Angeles International Airport showing its two 12,000-foot runways and two 10,000-foot runways (upper right). Pacific Ocean can be seen at upper left.

handed off by Los Angeles Center to Approach Control. Here's how it goes.

Pilot: Los Angeles Approach Control. This is United National 9 with you. Descending to one zero thousand and we have information Zulu. Over.

Controller: United National 9. This is Los Angeles Approach Control. Ident. After Arnold maintain seven thousand and two five zero knots. Over.

Pilot: United National 9. Roger. Maintain two five zero knots and seven thousand after Downey.

Controller: United National 9 at two five zero knots, cleared ILS runway 25L approach. Over.

Pilot: United National 9. Roger maintain two five zero knots, cleared ILS runway 25L approach.

Controller: United National 9. Reduce speed to one six zero knots and maintain one six zero knots to Lima. Contact the tower on one one eight point niner when over Lima.

Pilot: Reducing speed to one six zero knots and contact the tower on one one eight point niner over Lima.

Pilot: Los Angeles Tower. This is United National 9. Over Lima.

Controller: United National 9. This is Los Angeles Tower. Runway two five left, cleared to land. Wind two five zero degrees at one zero. Report the runway in sight.

Pilot: Roger. Report the runway in sight.

Pilot: Los Angeles Tower. This is United National 9. We have the runway in sight now.

Controller: Roger.

Observer: After we land on the runway, the tower controller will hand us off to Ground Control. The ground controller then gives us taxiway instructions, clearances to cross runways, and information as needed to lead us to our unloading gate.

Controller: United National 9. Turn left at the next intersection and contact Ground Control on one two one point seven.

Pilot: United National 9. Roger.

Pilot: Los Angeles Ground Control. This is United National 9. For the West Imperial Terminal.

Controller: United National 9, this is Los Angeles Ground Control. Taxi to West Imperial Terminal.

Captain: "Well, folks, here we are at the end of our trip. You have arrived safely and with little delay due to close cooperation between your pilot and your controller. Thus, you have seen at firsthand the teamwork between pilots and controllers which is so essential to effective functioning of the Air Traffic Control System. You have received some idea as to what takes place on the ground to safeguard your flight as it proceeds through the airspace. We have one of the finest ATC systems to be found anywhere in the world! But constantly increasing air traffic volume and more and more use of air transportation will require continuing refinements and improvements in the system. These are going forward on many fronts. Now that you have seen something of the exciting world of Air Traffic Control, perhaps you will want to learn more about what goes on 'behind the scenes.' I hope so."

CHAPTER II

HOW IT ALL BEGAN

7 *Typical control tower in mid-1930s.*

The control of air traffic began in attempts to "organize" flight operations through the application of rules, regulations, and procedures. These were supposed to be carried out by pilots and were, to a large extent, merely derivations of existing ground and marine "rules of the road." Since all flights were of short duration, and their completion as planned was somewhat doubtful, they were mainly of the "keep to the right" variety. Because of the relatively small number of aircraft, or "flying machines," there was no requirement for any form of ground-based control.

Early in the development of the airplane in Europe, it became apparent that at least some degree of standardization should exist. This was considered very desirable in order that aircraft could be flown in different countries under similar regulations. Because of this situation, nineteen nations of Europe tried to reach international agreement on such rules in 1910. Agreement, however, could not be reached at that time.

At the Versailles Peace Conference in 1919, following World War I, the International Convention for Air Navigation was agreed upon internationally. As a result of this, the International Commission for Air Navigation (ICAN) was created to develop, among other regulations, the "General Rules for Air Traffic."

Perhaps the "grandfather" of today's Instrument Flight Rules was the following ICAN admonition: "Every aircraft in a cloud, fog, mist or other condition of bad visibility shall proceed with caution, having careful regard to the existing circumstances." Since ICAN existed as the only international agency concerned with the operation of aircraft, its rules and procedures were applied in most countries where aircraft were operated.

In order to understand the manner in which these rules were implemented throughout the world, their development into a modern Air Traffic Control System will be examined in some detail, using the system in effect in the United States as an example. The development of air traffic control systems in other countries has followed a similar pattern, varying in details only, mainly due to differences in legislative procedure and technical developments.

EARLY UNITED STATES DEVELOPMENTS

Although the United States was not a signatory to the ICAN Convention, it followed some of these concepts when, in 1926, it developed a program to "establish air traffic rules for the navigation, protection and identification of aircraft, including rules as to safe altitudes of flight and rules for the prevention of collisions between vessels and aircraft." Commencing in 1927, a program was inaugurated to establish a "Federal Airways System," and a network of radio beacons and later a similar network of four-course low frequency radio ranges were laid out to connect principal cities in the United States. Included in this program was the installation of light beacons to assist in the identification of these airways at night. Radio communication between ground and aircraft was largely nonexistent until 1930, when two-way radio-telephone communication arrived. By 1932, virtually all airline aircraft were being equipped for radio-telephone communication with airline ground stations.

With the advent of two-way radio-telephone capability in aircraft, radio-equipped airport traffic control towers came into being to replace the "flag-men" up to then in use to control airport traffic at "busy" airports. The first radio-equipped control tower in the United States was put into operation at the Cleveland Municipal Airport in 1930, and about 20 other control towers were installed by the local governments in major cities during the next five years.

Up to 1933, aircraft were operating essentially under good weather conditions on a "see and be seen" basis. In that year, the United States government prescribed standards for proficiency in instrument flying and required that airline pilots demonstrate their qualifications in the "advanced art of flying." With the advent of instrument flying capability, it was just a short time until the problem of avoiding collisions while flying in bad weather became recognized.

During the latter part of 1934, the author, while employed by American Airlines as a radio operator/dispatcher, initiated a flight-following system for the aircraft of that company when approaching within about

8 Newark Airport Terminal Building housed the first airway traffic control center. Location was a room on the second floor, identified by the three windows in the wing on the right.

9 Newark Center under government auspices in 1936 was marked by first government-issued equipment—a large-faced clock.

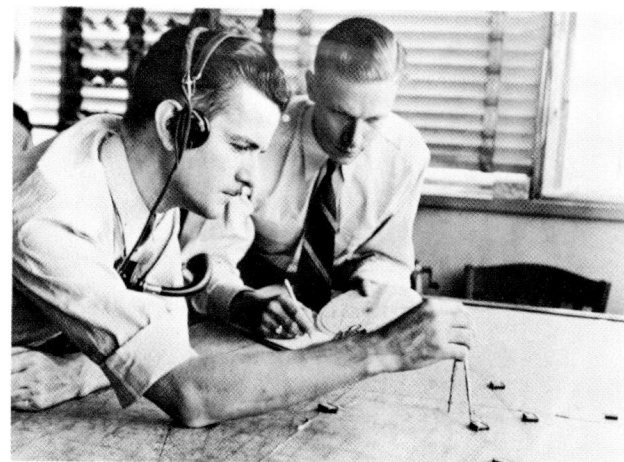

10 Moving "shrimp boats" on map table to follow flights in the "control area." (Author at right.)

a hundred miles of the Chicago area. Shortly thereafter other airlines flying into Chicago, such as United, TWA, Eastern and Northwest, cooperated in this informal effort. A similar coordination system subsequently was established in Newark — then the main airport serving the New York area — with the airlines using that airport. Up to this stage, in the event of any possible conflict with another aircraft, the dispatchers of each of the concerned airlines would coordinate arrivals with each other and the control tower by means of a local airport interphone.

FIRST AIRWAY TRAFFIC CONTROL CENTER

Realizing the importance of establishing a more formal coordination system, the principal airlines using the Chicago and Newark airports got together in 1935 to establish "Interline Agreements" for the purpose of providing a unified coordination mechanism to handle the airline traffic at each of these airports. It was then agreed that "Airway Traffic Control centers" (in later years to be called Air Route Traffic Control centers) would be established as soon as possible at Newark, Chicago, and Cleveland with all the airlines using these airports prorating the costs of each center according to their respective traffic volume. The first such center was established at Newark on December 1, 1935, and thus the world's first Airway Traffic Control Center came into being.

During the early months of 1936, the Newark Center served as a training ground for personnel scheduled to man the three centers under the cooperative plan of the airlines. In April of 1936, the second center was placed in operation in Chicago, and the third, at Cleveland, in June of that year. Each center "controlled" — through the airline radio operator/dispatcher — only airline aircraft (which comprised virtually all of the instrument traffic at that time) in an area roughly 50 miles from its terminal airport. The basic equipment in a center during this period consisted of a map table and a blackboard. Markers were used on a map table to indicate the position of each aircraft, while flight data were written on a blackboard. Later, paper flight-progress strips in movable holders replaced the blackboard, and use of the map table was discontinued.

During the interim period of airway traffic control operation by the airlines, steps were under way by the United States government to establish "a uniform and centralized system of Airway Traffic Control . . . to direct and coordinate the progress of all flights, whether government, civil or commercial, over the Federal Airways so as to insure the maximum safety in flight by preventing traffic confusion which might result in collisions, and to direct traffic so as to insure arrivals at airports in an orderly manner."

GOVERNMENT OPERATION

Thus, on July 6, 1936, the United States government assumed the operation of the three centers already in being, and during the following months established five additional centers. These centers were operated from 12 to 16 hours per day, and handled approximately 30,000 aircraft movements (mostly airline) during the first 12 months of the government's entrance into the field of airway traffic control.

Although the federal government assumed responsibility for providing en route or "airway" traffic control service, the airport traffic control towers continued to be operated by the local governmental authority which operated the airport on which the control tower was situated. There were no uniform standards for control tower personnel, and no comprehensive regulations governing control tower operation. Although a "Committee on Airport Traffic Control" had been formed in 1929 by the then Aeronautics Branch of the Department of Commerce, it had no authority in the matter. In its final report published and released in 1933, it said in part that "at airports where traffic movements are infrequent or sufficiently spaced, experience has demonstrated that the Uniform Field Rules supplementing the Federal Air Traffic Rules are sufficient to preserve order in the traffic flow. However, traffic densities at many of our airports have already reached a point where definite control, involving signalling equipment of various types and a carefully thought out plan of traffic flow is necessary. With the continued development of air transportation and the consequent multiplication of traffic density, the need for

11   Contact flight rules (CFR), later to be called visual flight rules (VFR), have remained essentially unchanged through the years. (Illustration from the author's book, Air Traffic Control, *published in 1945.)*

12   Air route traffic control center in the early 1940s using movable flight-progress strips in lieu of old blackboard postings.

such control will be still more pronounced."

During the formative years of the Airway Traffic Control System, valuable counsel and advice were given by such organizations as the Air Line Pilots Association (founded in 1931); the Air Transport Association, representing the airlines (founded in 1936); the Radio Technical Commission for Aeronautics, a nonprofit cooperative government/industry consultative organization (founded in 1935); Aeronautical Radio, Inc., a communications operation arm of the air transport industry (founded in 1929); and the Aircraft Owners & Pilots Association representing a large segment of private and business pilots (founded in 1939). The first government/industry coordinating group formed to assist in developing the Airway Traffic Control Service was an "Airways Operation Advisory Committee" created in 1937, the industry representatives on which came largely from the above-mentioned organizations. Later on, this consultative concept was expanded by forming an "Air Coordinating Committee" consisting of more comprehensive government and industry representation.

REGULATORY ACTIONS

A major step forward in modernizing aviation legislation in the United States was the Civil Aeronautics Act of 1938. This act provided for a new regulatory code — the "Civil Air Regulations" — of which CAR 60 (Air Traffic Rules) formed a significant part. Also created by this act was a new aviation agency of the United States government — the Civil Aeronautics Authority (CAA) which included the Airway Traffic Control Service. Subsequently, the operating functions of the CAA were included under the Department of Commerce as the "Civil Aeronautics Administration," and the economic functions in a new "Civil Aeronautics Board."

In 1958 a new "Federal Aviation Act" was passed by the United States Congress, creating an independent Federal Aviation Agency (FAA) as the successor to the CAA. At this time the Federal Air Regulations (FAR) replaced the CAR. A significant principle of this act was to strengthen the previously established concept that the federal airways system should be a "single common system designed for and used by, both civil and military aviation." On March 1, 1967, the Department of Transportation came into being and included the FAA among its numerous agencies and bureaus. At that time, the name of the aviation agency became the Federal Aviation Administration, although its functions and responsibilities remained virtually unchanged.

A significant part of the CARs of 1938 was to establish a set of air traffic rules, more precisely defining contact flight rules (CFR) and instrument flight rules (IFR). For flight under IFR weather conditions, the pilot was required to have an instrument rating and the aircraft was required to be equipped with a minimum of certain instruments, most notably two-way radio. The pilot was required for the first time to comply with instructions issued by an airway traffic control center; up to this point, traffic control instructions had been merely advisory.

Included in the CAR was the establishment of a Civil Airways System (green, red, amber airways delineated flight levels when not under Airway Traffic Control jurisdiction); designation of controlled airports; airway traffic control areas; and radio fixes over which position reporting was required under IFR.

EXPANSION PROGRAM

With the increased recognition and authority of airway traffic control centers under the new Civil Air Regulations, a program of expansion commenced in the late 1930s. This included the installation of teletype communications between the centers and the aeronautical communications stations (most of the radio range stations by then had voice communication capability), supplemented by more interphone communication with outlying airports and airline dispatchers. With this expanded communications service, it was possible for the first time to actually control all airway traffic under IFR within a center's control area. By mid-1938, eight airway traffic control centers were in 24-hour operation, controlling about 26 percent of the designated airways.

The next three years saw significant advancements in the Airway Traffic Control System. Six additional centers

13 Forerunner of digital data processing was a display demonstrated in the Washington Center during 1940-1941 using revolving drums. Flight-plan data were inserted by controller operated input devices (keyboards).

14 Updated control tower equipment after government operation was assumed, marked by introduction of flight progress boards similar to those used in the centers.

were activated and plans were well underway to provide complete airway traffic control service for the entire United States. The teletype network was further expanded, supplemented by a complete national interphone system connecting centers, communications stations, airport control towers, weather facilities, and military bases. With the increased communications service, the efficiency of the Airway Traffic Control System also increased at a corresponding rate.

An interesting forerunner of the modern automation program was the installation in 1940 of experimental automatic flight data posting equipment in the Washington Airway Traffic Control Center. This equipment — in those days sometimes called "the mechanical monster" — was, in effect, an automated flight data processing prototype. World War II, however, stopped the further development of this approach, and an automation program for the control of air traffic was not renewed until the late 1950s.

Another device in the development stage at this time in the United States was an airborne collision warning indicator, tested as a "breadboard" model during 1941. But this project also was dropped as a result of the war, not to be reactivated on a renewed experimental/testing basis until the 1960s.

At the end of 1941 there were approximately 300 personnel employed in 14 Airway Traffic Control centers covering some 20,000 miles of airways in the continental United States (about 54 percent of the total number of designated airways). During the first part of 1942 nine additional centers were commissioned, giving 100 percent coverage of the airways system for the first time. In addition, a coordinated effort was inaugurated to provide CAA Airway Traffic Control personnel at certain military facilities to make coordination of civil/military air traffic control more efficient.

By 1941, there were about 150 airport traffic control personnel in the United States employed in control towers which, up to that time, had continued to be operated exclusively by municipal and other local governments. Under the provisions of the Civil Aeronautics Act of 1938, all such personnel had merely been licensed as to proficiency by the federal government. Due to the war, however, the limited number of towers in operation failed to meet the requirements of the military services and the flying public and a program was undertaken to extend air traffic control under the federal government's jurisdiction to include Airport Traffic Control. Thus, the CAA in November of 1941 for the first time entered into a coordinated system operation including both airway traffic centers and airport traffic control towers, and the United States' "Air Traffic Control [ATC] Service" came into being.

In anticipation of the need to expand the ATC facilities rapidly and considering the likelihood of United States involvement in World War II, a comprehensive training program was launched during 1941. Seven training centers were established in 1941-1942 in key locations within the country, and a massive recruitment drive was initiated. Women for the first time were brought into the Air Traffic Control Service, and at exactly the same salaries as were paid to men for equivalent posts. (By the end of the war, approximately one-third of the controller work force would be women.)

With the consolidation of all air traffic control functions, the jurisdiction of the airport traffic control towers was extended beyond local landing and takeoff control to include "approach control" under IFR conditions. This was made possible by coordination of authority between the centers and the towers, and permitted delegation of responsibility, hitherto held by the centers, to the individual control towers.

The airway traffic control and airport traffic control facilities became completely integrated during this period, and the level of efficiency of control techniques was constantly raised. The net result was the evolution of a true "air traffic controller profession." By 1946 the CAA was operating 113 control towers and 24 centers employing a total of approximately 1,800 controllers. The experience gained in the United States during this first ten-year period set the stage for the continued expansion of the proven ATC System with improvements in procedures, equipment, and controller training.

Prior to and during World War II, the Air Traffic Con-

During the early years of air traffic control and on into the 1950s, low-frequency radio ranges provided navigation guidance along four courses. The pilot followed these "beams" by flying his aircraft in the zone between the "A" and "N" signals, identified as a steady hum in his radio receiver. In addition to providing en route navigation, these radio ranges were used for instrument approaches to a nearby airport.

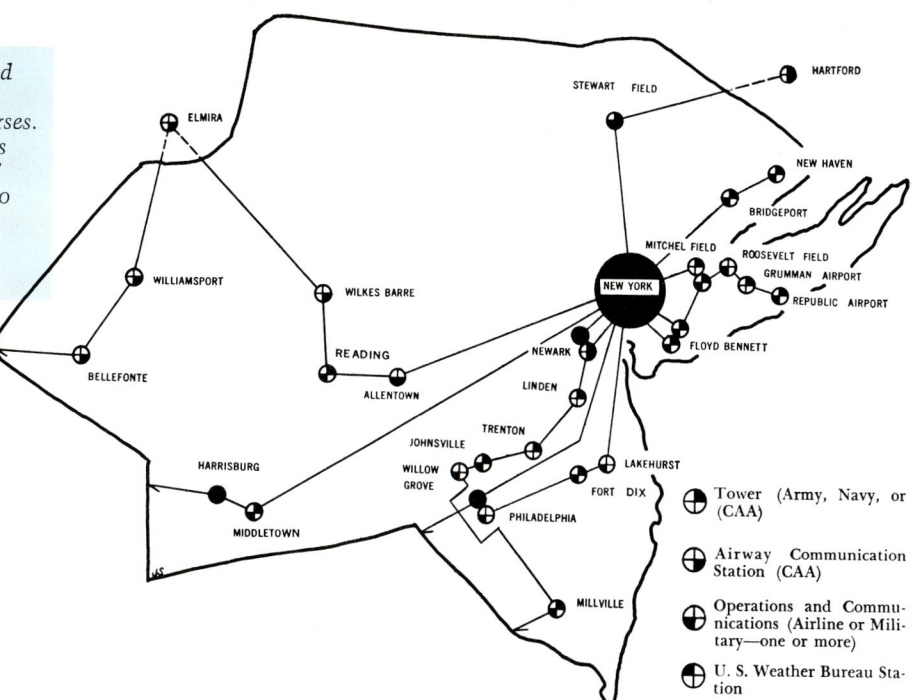

15   First comprehensive communication linkage between a center and other facilities within its control area.

16   General Aviation pilots being briefed at a flight service station.

trol Systems of the United States and Canada were using almost identical procedures and equipment. Cooperation between various countries in Europe along similar lines had to be postponed until after the war. In other countries, the volume of air traffic was not sufficient to warrant the implementation of such a system at that time.

COMMUNICATION DEVELOPMENT

During the first fifteen years of operation, the Air Traffic Control System used indirect methods of communication between controllers and pilots — except in the limited local communication area of a control tower. Communications with airline aircraft were carried out by relay via the radio communication channels of each airline company: controller-to-dispatcher-to-pilot-to-dispatcher-to-controller. Non-airline aircraft (military and general aviation) communicated with "Airway Communication stations" associated with the four-course low frequency radio ranges used by all classes of air traffic for navigation. Again, the communication link between controller and pilot of these categories of aircraft was indirect as was the case of airline aircraft. The personnel at such a basically navigation station, subsequently to be called a flight service station (FSS), gradually assumed more functions for expanded flight-information dissemination to pilots.

By the 1960s the FSS facilities became oriented to serve general aviation in flight planning by making national weather service information directly available (airlines and the military continued to use their own operations and dispatching facilities). With the advent of the 1970 decade, flight service stations were providing many services to general aviation pilots, such as preflight briefing, en route communication with VFR flights, lost-aircraft assistance, notices to airmen (NOTAM's), weather broadcasts, and accepting and closing all classes of general aviation flight plans. The FSS became tied in directly with the appropriate ATC facilities for IFR flight-plan transmission and for IFR ATC clearance delivery at airports not having a control tower. Long-range plans would call for FSS facilities at a sufficient number of airports by the late 1970s to enable serving directly all of the United States' general-aviation traffic, including international FSS service.

During the early 1950s, remote communications air-ground (RCAG) facilities were developed to provide direct pilot/controller communications for all classes of IFR traffic in order to eliminate the laborious and inefficient indirect system in use up to that time. Although ATC communications were removed from airline radio channels as a result of the RCAG program, the airlines continued to develop and expand their own communication facilities (for non-ATC purposes) under the auspices of the airline-owned Aeronautical Radio, Inc. (ARINC).

INTERNATIONAL IMPACT

With the end of World War II approaching, an International Civil Aviation Convention was held in Chicago, attended by representatives of 52 nations, which set the stage in the latter part of 1944 for the international development of civil aviation in the postwar era. Here, the foundation was laid for the various annexes to the convention, including one on "Air Traffic Control" and another on "Rules of the Air." (The experience of the United States Air Traffic Control Service contributed significantly in the formulation of these annexes.). A Provisional International Civil Aviation Organization (PICAO) was established to develop agreements between member nations for the promulgation of appropriate standards and recommended practices, as well as to take actions which would promote the implementation of these standards all over the world. The initial efforts of PICAO were taken over by the permanent International Civil Aviation Organization (ICAO) on April 4, 1947.

The standards and procedures which have been developed by ICAO for the control of air traffic (a) apply directly and solely over the high seas, and (b) are followed by member countries through incorporation in their national regulations for flight over their respective airspace. For example, in the United States, the Federal Aviation Administration (FAA) has the responsibility and statutory authority to implement the international standards and procedures for all flights over its territory.

17 First major independent control tower structure at New York's Idlewild Airport (subsequently JFK International) contained all of the expanded functions for local and terminal area control facilities.

18 The pioneer air traffic controllers in the early days and 25 years later.

As it became necessary to develop special procedures in certain areas of the world, ICAO regions were established and regional meetings attended by interested governments were held to agree on such procedures. An air navigation plan for Air Traffic Control services was established in each of these regions, and these are continuously updated under the auspices of the permanently structured ICAO regional offices.

Since the creation of ICAO, worldwide development of uniform air traffic control services has taken place in many parts of the world, stemming from the ICAO standards and regional arrangements. Some of these developments have resulted also from direct cooperation between countries (such as the close coordination of the Canadian and United States air traffic control operations, especially over the North Atlantic, where the United Kingdom and Ireland also work very closely together; and Eurocontrol, handling the air traffic over the territories of Western European countries.)

The advent of the subsonic jet aircraft in the late 1950s, coupled with the introduction of supersonic transports (SST) and "jumbo jets," bring into focus even more the international nature of air traffic control. Today, aircraft of many nationalities operate over the territories of virtually all countries of the world and are being provided more and more the services of uniform and equally efficient air traffic control. This cooperation must continue and in fact be expanded so that eventually there will be no geographic or technical "gaps" in the world's ATC System.

CONTROLLER PROFESSION

With the recognition of the importance of Air Traffic Control Service internationally, there also has come about international recognition of the professional status of the air traffic controller. Air traffic control associations have been formed in many countries, the members of which are professional controllers in each such country. These associations generally have as their objective the voicing of controllers' interests in terms of human considerations such as working hours, salaries, and employment conditions. Additionally, they also are playing a role which should become more significant in the future — in expressing the collective viewpoints of controllers on such fundamental questions as air traffic control equipment, procedures, and technological development needs.

A significant step in the direction of multinational professional controller representation was the creation in 1961 of the International Federation of Air Traffic Controllers Associations, (IFATCA) with member associations from all over the world. Their contribution will become increasingly important in the continuing development of an international Air Traffic Control System which will serve most effectively and efficiently the world's air transportation needs.

CHAPTER III

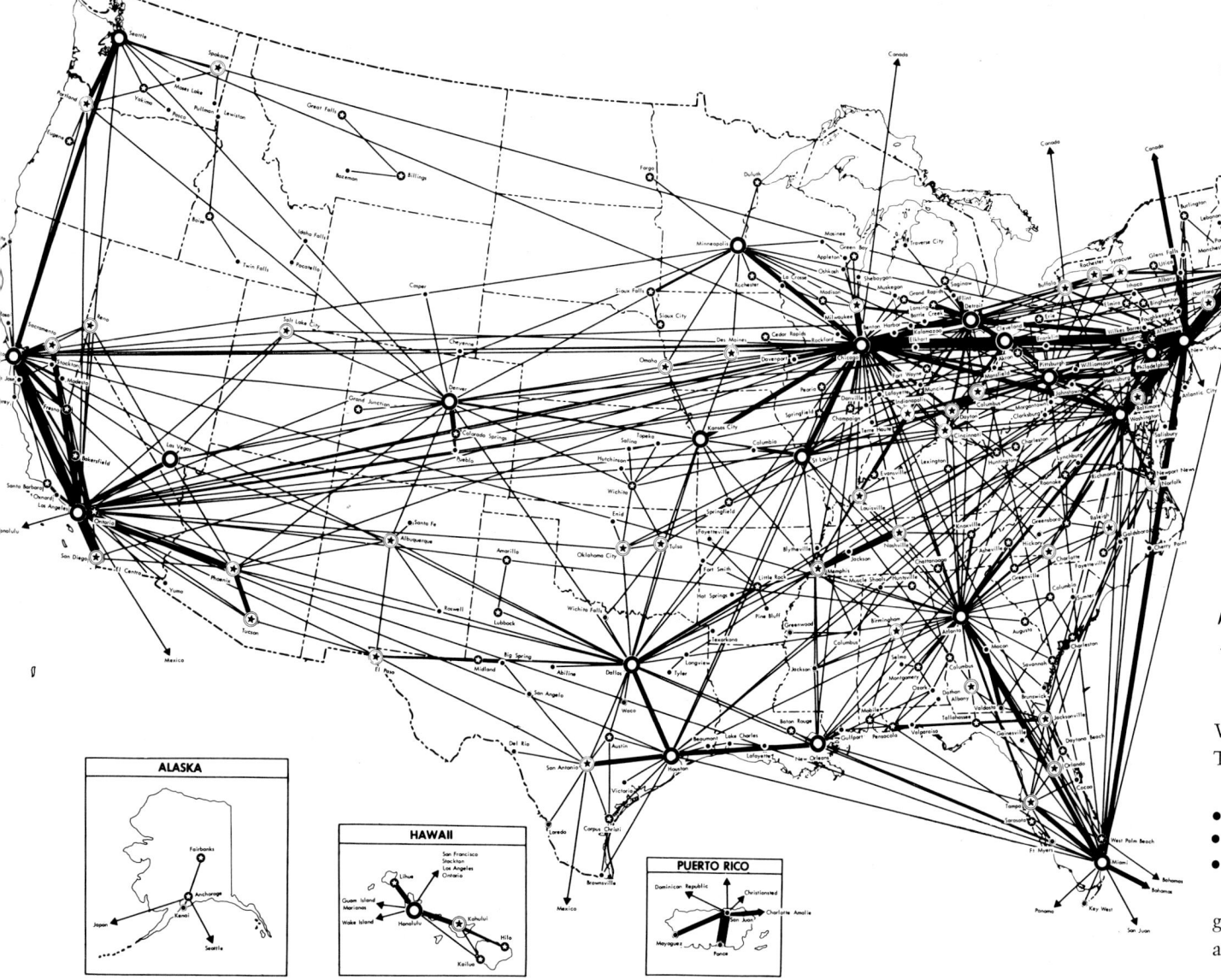

19 *The major traffic patterns in the United States.*

THE USERS

Who are the users of the airspace, and in turn the Air Traffic Control System?

They are classified most basically as:

- Air carriers
- General aviation
- Military

Each of these broadly defined users contains subcategories identified by different types of transportation applications.

Also, there are cross-classifications related to aircraft characteristics and operational considerations.

All of these variables directly affect the demands placed on the ATC System. Thus, an acquaintanceship with "the users" — who they are and what they do — provides important background in understanding the significance of the previously expressed fundamental challenge which the ATC System must meet "to provide the means which will make possible the safe and expeditious use of the airspace by all who desire to use it as a transportation medium."

AIR CARRIERS

These users provide scheduled (or chartered) service to the general public for hire. They want to meet their schedules *on time* at their predetermined arrival points. In general, they follow the instrument flight rules at all times, regardless of weather, under positive control within the ATC system. They have three basic subcategories — trunk, regional, and short haul.

20 *Representative airplane used by trunk air carriers.*

21 *Typical aircraft used by regional air carriers.*

22 *Aircraft used in short-haul air transportation.*

Trunk

The trunk air carriers essentially operate the long-range, multiengine, large jet aircraft (125+ passenger capacity) on stage lengths generally in the order of 1,000 miles or more. As a rule, they operate into the larger airports serving the more populous metropolitan areas. Their typical flight time per trip is two to four hours.

Regional

These air carriers serve medium-size communities with stage lengths in the order of 500–1,000 miles. Their approximate flight time per trip is one to two hours. They use medium-size aircraft (50 to 100 passenger capacity) and operate into medium-size airports in general, while at the same time serving the same airports as are used by the trunk carriers.

Short Haul

The local, or "short-haul" air service as it is more commonly referred to, satisfies both *metropolitan* and *rural* transportation requirements. Typical stage lengths run from fifteen minutes to perhaps one hour, utilizing aircraft with widely varying passenger capacities and speeds. In metropolitan areas, short-haul service operates in the following air transportation modes.

• Intercity mode, in which the short-haul service connects cities when the intercity distances are on the order of 500 miles or less. The short-haul service may operate into an existing city-associated jetport and/or a central business-district landing facility. Examples are Portland, Oregon, to Seattle, Washington; Boston, Massachusetts, to New York City.

• Intraurban mode, in which the short-haul service operates between urban centers that sprawl about city centers as well as serving city-center landing facilities and city-associated jetports; for example, to and from the urban areas surrounding New York City and interconnecting with Kennedy, La Guardia, and Newark jetports, plus city-center facilities such as the Wall Street heliport.

• Regional jetport mode, in which the short-haul service carries passengers to and from a regional jetport from airports situated in urban sites in a region; for example, between cities comprising the greater Los Angeles area and the Palmdale jetport.

In rural areas, the short-haul service functions in the following air transportation modes.

• Intrastate mode, in which the short-haul service operates between distant reaches of a state or collection of states into the central business area; for example, from western Massachusetts and points in New England into Boston.

• Recreational mode, in which the short-haul service feeds recreational centers from urban points; for example, from Denver to the ski areas in Colorado.

• Natural resources mode, in which men and supplies are transported to remote areas; for example, from shore bases to offshore-oil drilling rigs.

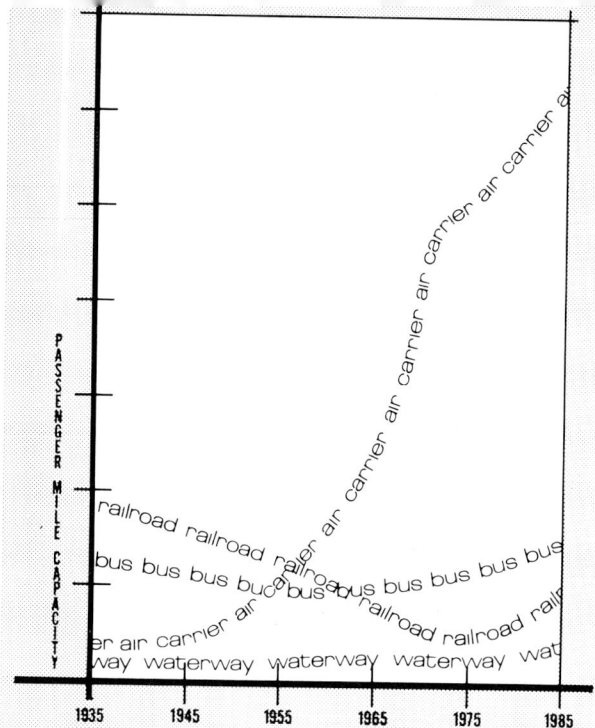
23 Passenger-mile comparative capacity of the different types of common carriers.

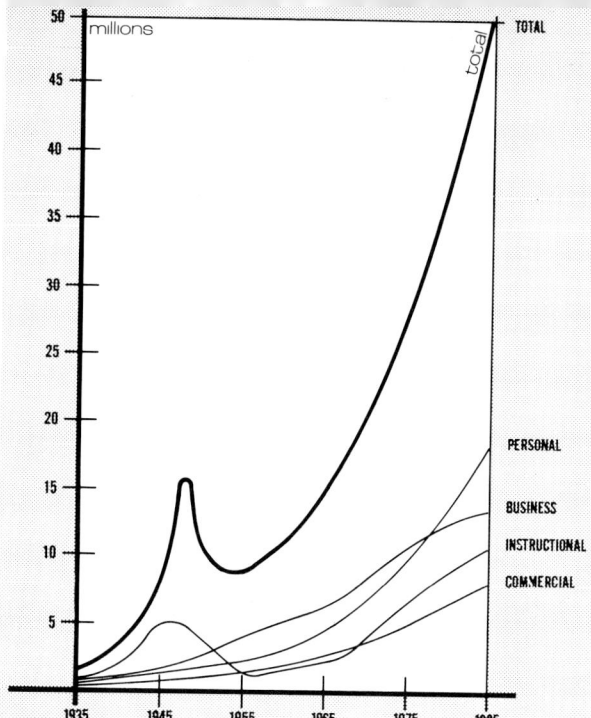
24 Distribution of different general aviation activities by flying hours.

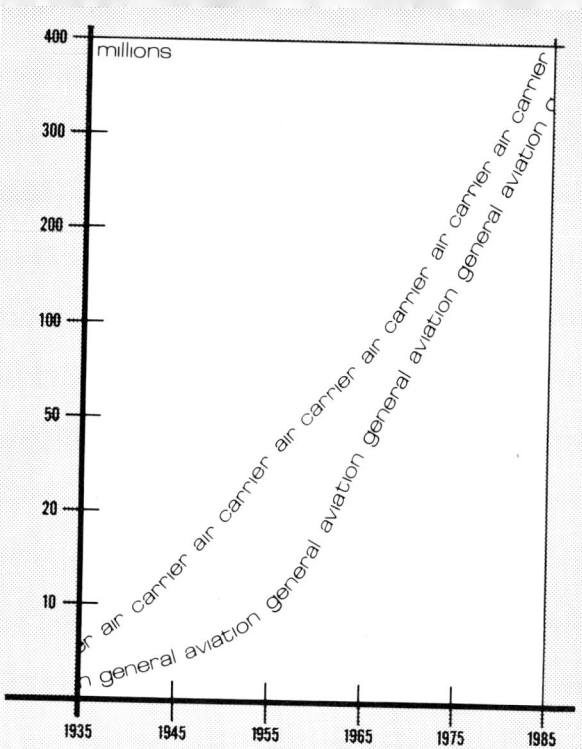
25 Comparison of passengers carried by air carriers and general aviation.

26 Representative aircraft used in personal, corporate and business flying.

GENERAL AVIATION

General aviation is comprised of several distinct categories covering the following.
- Personal flying, including flights performed for pleasure and nonbusiness purposes.
- Business flying, including flights carried out by companies and individuals in the course of conducting their particular business.
- Commercial flying, including such activities as air taxis, aerial application of insecticides, pipeline patrol, aerial surveys, aerial photography, and police surveillance.
- Instructional flying, including flights performed in the course of training pilots.

Aircraft have been used for personal flying all over the world ever since man learned to fly. On a worldwide basis, it is probably the largest segment of general aviation, and is constantly growing at a rapid pace. The other categories, which could be grouped in a general way as "nonpleasure," also account for a significant portion of general aviation's total activity.

In some countries, as for example in the United States, general aviation has made swift progress, due, in part, to encouragement and lack of unduly restrictive regulations by the government. In other countries, however, military and airline operations have received priority treatment to the detriment of general aviation's development. While the sheer number of general aviation aircraft reflects a significant demand upon the ATC System, it is also true that many of these aircraft conduct most of their operations outside of the system where no type of air traffic service is needed. The number of general aviation aircraft handled by the ATC System, however, is constantly moving upward, particularly in the substantially increasing volume of such aircraft which are operated under positive air traffic control procedures (IFR).

The nature of the ATC Service provided to the pilots of these aircraft is almost identical to that provided to the other categories. An IFR flight made by a general aviation aircraft is given the same handling by ATC facilities as are the scheduled air carriers, for example. Since many such flights operate into and out of airports not served by the air carriers, slight differences in handling, as well as in formal procedures, do arise occasionally. Such differences result mainly from the different degrees of skill, training, and experience which apply to some classes of the general aviation pilot.

These classes would include naturally those who are still in the training stage, as well as those who fly exclusively for a hobby in good weather. On the other hand, general aviation also includes professional pilots, in the case of business flying for example, who may be part of a highly organized department of a large corporation. Those pilots have the skill and training equal to an air carrier pilot as well as a wealth of experience in all types of aircraft including those used by the air carriers.

27 Military Air Lift Command Hercules is typical of military transport aircraft flying within the Air Traffic Control System.

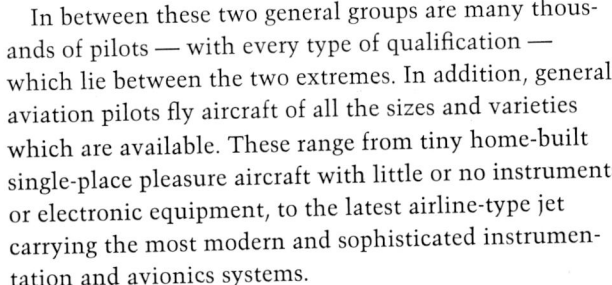

28 Numbers of aircraft by airspace users.

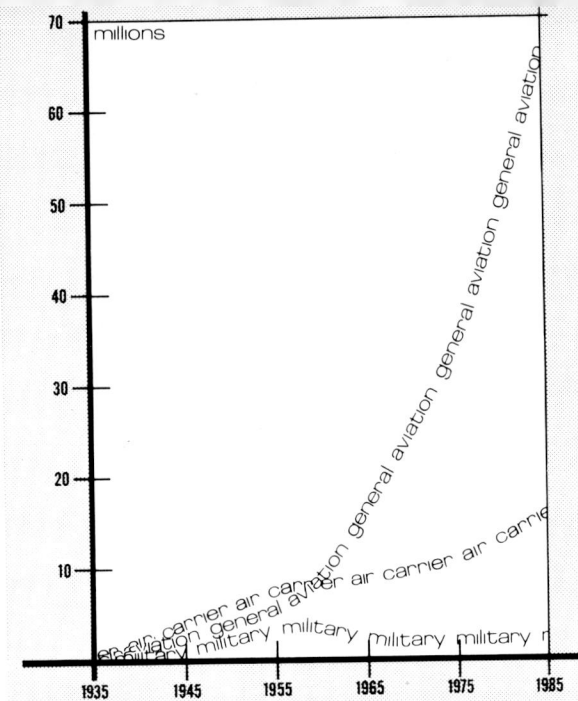

29 Annual flying hours performed by the airspace users.

In between these two general groups are many thousands of pilots — with every type of qualification — which lie between the two extremes. In addition, general aviation pilots fly aircraft of all the sizes and varieties which are available. These range from tiny home-built single-place pleasure aircraft with little or no instruments or electronic equipment, to the latest airline-type jet carrying the most modern and sophisticated instrumentation and avionics systems.

MILITARY

The effect that military operations have on ATC Systems over the world is extremely varied. Among the many reasons for this is that most such operations must receive the highest priority. As a result, some nations have set aside designated areas of their airspace for the sole use of military aircraft. Such areas may comprise specially defined sections of the country from the surface upward to infinity. Other such areas may be reserved for certain specified periods and between specified altitudes. A variation to this form of reservation is one in which all the airspace between specified altitudes may be prohibited to all except military aircraft. In other cases, blocks of airspace are reserved for a specific mission and only the minimum amount of airspace is used.

The military services generally provide their own air traffic control service within these reserved areas and also at their own airports. Military air traffic controllers in the United States meet qualifications similar to their civilian colleagues, and they function in close coordination with and as part of the overall ATC System. At some military airports having mixed civil-military air traffic, control may be performed by civilian controllers. Other than these limited instances, all military aircraft flying over United States territory (or in oceanic control areas under the jurisdiction of the United States) are controlled by the FAA's civilian Air Traffic Control Service and in accordance with the federal aviation regulations.

When certain military tactical operations take place, the ATC Service may authorize their operation through civil traffic airspace, although they may make their departure from and approach to a military airport. Special procedures provide, for example, for interception aircraft to make their climb or descent within the limits of an airway, if necessary.

The volume of military air traffic fluctuates according to the political situation at any given location and time. Large numbers of such fleets can, under modern tactical concepts, reach any spot on the earth in a very short time. This capability, in many cases, will permit an overall reduction in the number of aircraft assigned to an individual military command.

A form of military airspace reservation which is used extensively in the United States is one which is specially designed for a particular mission. A central altitude reservation facility (CARF) is operated by the FAA with the full responsibility to "coordinate and to approve altitude reservations in the airspace of the Continental United States and its possessions and to deal directly with the ATC facilities of foreign countries." Some of the military missions which are coordinated and approved by CARF and for which altitude reservations are made include:

- Presidential flights
- Hurricane evacuation flights
- Large-scale exercises
- Mass movements of fighter aircraft
- Official speed tests
- Routine training missions
- Military airlift command flights
- Satellite and missile launches.

Each altitude reservation normally includes the airspace of several FAA air traffic control facilities and thus requires the cooperation of the controllers in each such facility. In addition, altitude reservations include the departure, climb, cruise, and arrival phases of flight, up to the arrival holding pattern at which normal separation standards are placed into effect. These altitude reservations are normally employed when a number of aircraft must be moved with less than standard separation.

It is the responsibility of CARF to coordinate requests for altitude reservations with the appropriate United States and foreign ATC facilities. Each such facility indicates its approval for the altitude reservation by so ad-

CHAPTER IV

THE HUMAN ELEMENT

vising CARF. In this manner, the vital military traffic can be successfully assimilated into the overall ATC System without undue difficulty.

USER NEEDS

All of the different users of the ATC System have essentially the same needs.
- To avoid midair collisions and other possible hazards to safe flight.
- To use the airspace and airports with maximum flexibility and minimum restrictions.
- To be able to fly with little or no delay due to air traffic control.

While simple to express, these user needs involve considerable effort to implement as air traffic volume continues to increase. The distinctive operational requirements and characteristics of the different user categories pose a significant challenge to the ATC System planners and operators if user needs are to be met effectively.

Subsquent chapters of this book will discuss and analyze the various elements comprising the Air Traffic Control System in relation to these users' needs.

To work effectively, the Air Traffic Control System must include an adequate number of efficient human beings — just as any other complex "system" does. The ATC System, however, has a requirement for more people having varied capabilities than many other systems. Teamwork must receive much more emphasis in this area, due to the system reliance upon close coordination of the actions of many people doing many things — all aimed towards the attainment of the overall goal of meeting most effectively the users' needs.

While the pilot/controller team is well recognized and at least fairly well understood, there are many other humans who are involved in the day-to-day operation of the ATC System.

Such people as those who maintain the equipment — both airborne and on the ground — must be well trained and efficient technicians as part of the system. Communications personnel, both government and private, who provide a link between the pilot and the controller also play a vital part of the system, and must maintain their efficiency and skills. Weather observers and forecasters are very important elements in the system, as well as personnel (such as those in a United States flight service station) who provide briefing and special communications services to pilots. Airline dispatchers and military operations personnel are among others who contribute significantly to the overall ATC System. While all of these well-trained people are essential elements of the system,

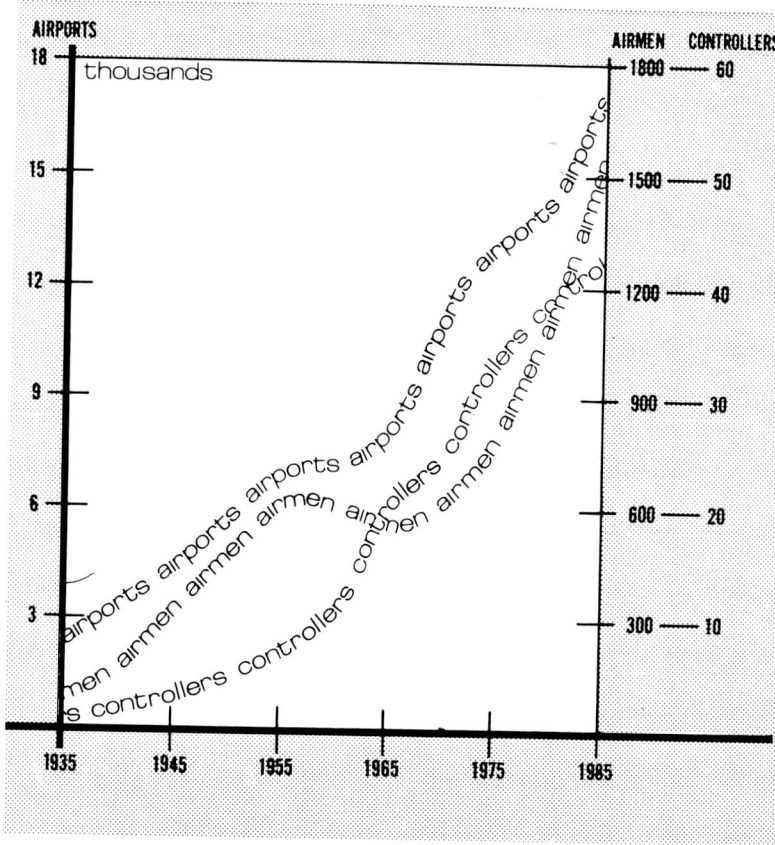

30 *Airmen, controller, and airport interrelationship.*

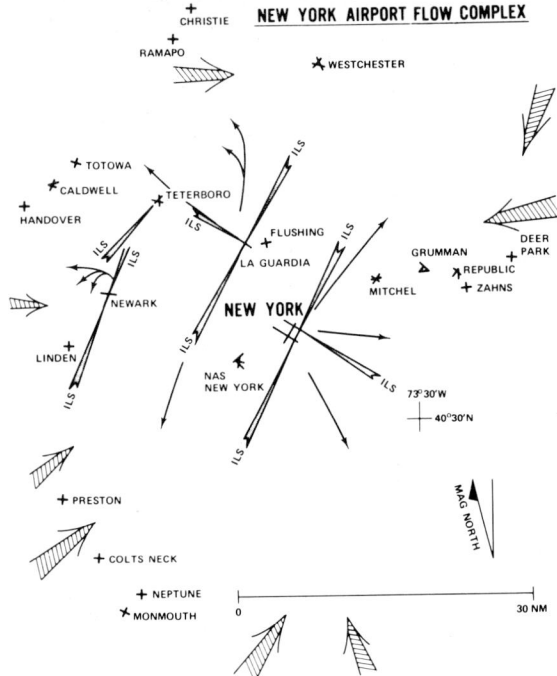
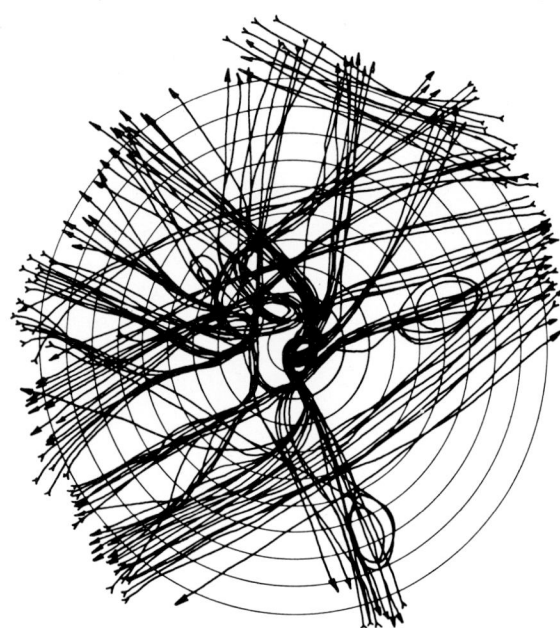

31 Complex traffic-flow patterns require close coordination between controller and pilot to achieve efficient use of airspace and airports.

32 Composite traffic-flow patterns in the New York area in one hour reflect the high degree to which pilot/controller cooperation is required.

33 Intense concentration on traffic movements is a part of the controller's life.

without whom it could not function, the human beings who are most directly involved in its safe and efficient operation are the pilot and the controller.

The pilot, in addition to operating his aircraft in accordance with the appropriate rules and regulations, must be fully cognizant of the requirements of the ATC System. These requirements include not only that he must thoroughly understand the various air traffic rules and regulations and control instructions that are issued to him, but he must also be proficient in carrying them out in the proper manner. To do otherwise would not only endanger his aircraft, but also other aircraft in the vicinity. In fact, the degree of efficiency attained by a pilot in his operation within the ATC System determines to a significant extent the efficiency of his overall performance as a pilot. It is obvious that pilot competency and cooperation within the ATC System are, in turn, factors which directly affect the overall efficiency of the system.

CONTROLLER PREPARATION

It is only natural, perhaps, that the air traffic controller should be considered as the focal point in the ATC System. While the situation varies quite considerably over the world, controllers generally begin their careers while they are relatively young, with an average starting age from 22 to 26 years. They usually select this career because of a knowledge of, or interest in, some facet of the aviation industry. A great many are civilian pilots, or have had service in military aircraft as pilots, radio operators, or navigators.

Because of the rapidly expanding requirement for the provision of additional air traffic control services, however, it may be necessary in some instances to recruit controllers from other sources. It is considered desirable that personnel who are to become controllers should have at least some college background, and preferably a college degree, or show equivalency in intelligence and aptitudes. They also must meet physical requirements generally equal to those of an airline pilot.

Training methods vary quite considerably throughout the world. In some cases, formal ATC schools or "academies" are established, complete with all forms of training facilities, including such things as model airports and full-size control towers. One example of this approach to formal training is that employed in the United States by the Federal Aviation Administration at its Aeronautical Center in Oklahoma City. The objective of this training program is to graduate a sufficient number of junior level controllers each year to meet increased controller requirements in the various ATC facilities due to growing traffic volume as well as to fill vacancies caused by normal attrition. These new junior controllers are specially trained for service in either tower or center facilities. Special courses for training personnel for the supporting flight service stations also are conducted.

Tower controllers are trained in a simulated airport traffic control tower with approach, departure, and ground control facilities. They receive courses in all subjects relating to the operation position for which they have been selected. Center controller training takes place in a simulated air route (area) traffic control center, including standard consoles, flight data boards, and communication hookups. In both cases, training is centered on basic procedural control. Radar and more advanced procedures are taught after the controller-trainee has acquired some experience in an operating facility.

The courses in the Aeronautical Center take from three to four months, with an additional 30 to 45 days of training given at the facility to which the graduate trainee has been appointed. A total period of from two to three years normally is required to qualify as a full-fledged controller.

In addition to nationally operated formal training facilities such as those in the United States, there are a number of international training centers in various parts of the world operated under the auspices of the International Civil Aviation Organization. These centers serve several countries within their region.

Another method of training controllers is through the use of nongovernmental organizations. In such cases, the company operating the training facility undertakes the training on behalf of a government, which then provides ATC Service as required. This type of training may follow any of the methods described, according to the wishes of

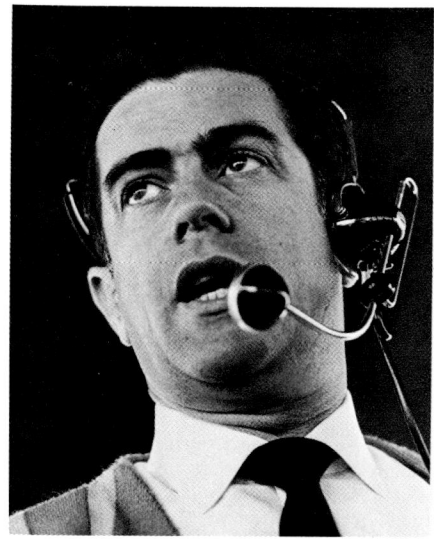

34 *Confidence in the decisions made on split-second notice is an essential ingredient in the controller's makeup.*

35 *Coordination and cooperation between controllers is a requirement for efficient handling of air traffic flow.*

36 *Canadian controllers have the same pressures and workload as their counterparts in the United States.*

the government concerned. In other cases, complete on-the-job training is done at the ATC facility in which the trainee will operate as a controller when his training has been accomplished.

CONTROLLER FUNCTIONS

Upon completion of his training period, and after successfully passing the appropriate examinations, the trainee is awarded a certificate or license by the appropriate governmental agency. In most cases, such a certificate authorizes the controller to provide ATC Service in a certain type or class of ATC unit. As experience is gained, additional authorizations may be given which, in turn, provide a basis for promotion of the controller.

A "journeyman" controller — one who is fully qualified to control traffic on his own responsibility — may be detailed to any one of several positions (discretely defined functional work assignments) within a particular tower or center according to his qualifications and the immediate work distribution requirements. He may occupy two or more such positions within that facility during the course of a day's work.

In a typical center, for example, there can be a number of positions which control traffic by means of a radar display, and other positions which control traffic in areas where there is no radar coverage. Others will be responsible for receiving flight plans from outlying airports, airline offices, or military bases, from which they prepare flight data reports, either manually or by means of a computer — if one is available. In some cases, the duties performed at certain positions are further distributed among subpositions or "sectors."

Towers also have a number of positions. Depending upon volume of traffic and complexity of the airport, there can be a "local" controller who controls the aircraft during takeoff and landing; a "ground" controller who is responsible for the aircraft while on the ground; "approach" and "departure" controllers; and a separate position which transmits only en route clearances (from the center) to the pilots of departing aircraft. These various tower positions may be subdivided into sectors as in the case with centers.

CONTROLLER PRESSURES

While the tremendous increase in the number of aircraft handled by the ATC System has meant expansion and a resultant greater employment of controllers, it also has developed additional problems for the controller. When aircraft speeds were relatively slow, and the number of aircraft using air traffic service was considerably fewer, a controller had a reasonable amount of time in which to make control decisions. With constantly increasing traffic volume and aircraft speeds, more and more control decisions have to be made in less and less time. The provision of additional controllers only alleviates the problem, it does not cure it. After a certain number of addi-

37 The pressure is on.

38 Calmness under pressure goes a long way.

tional controllers take up their positions within an ATC facility, the difficult problem of coordination between controller positions arises. To a large extent, this extra, coordination workload can lead to diminishing returns as the controller staff is increased.

A high percentage of the controller's workload results from his issuance of instructions to pilots (via direct communication radio links) as to heading, track, altitude, speed, and other flight parameters which the pilot must execute. These instructions provide the basis for establishing separation between aircraft for collision-avoidance purposes, as well as to meter traffic flow into and out of airports as required to adjust traffic volume to airport capacity. Other controller/pilot communications include such things as information on weather, other traffic and airport conditions. Included in the controller's communications workload are the corollary replies and verifications by the pilot of the control instructions and information. At the same time, the controller must devote close attention to the constantly changing traffic data shown on radar and other displays, and make appropriate split-second control decisions.

Because of constantly increasing pressures, the controller very frequently finds that his health, both mental and physical, has become adversely affected. This, in turn, frequently results in a lessening of his productivity in terms of the number of aircraft which he can handle efficiently. In some cases, a controller may "slow down" the traffic flow so that his personal capacity is not exceeded. As a result, delays to air traffic become more frequent and lengthy, which can place further psychological pressures on the controller.

Another facet which adds to the controller's pressures evolves from the fear that he may become a party to a legal action arising from an aircraft accident in which one of his control decisions may become a factor. The most serious type of such an accident is, of course, the midair collision. In such a situation, the legal authorities may consider that the controller's judgment might have been faulty and a wrong instruction was issued, or that he may have neglected to issue an instruction when an appropriate one might have prevented the accident. There are, of course, numerous variations to such situations, but these examples illustrate the type of legal difficulty which can so vitally concern controllers.

To alleviate the communications workload and related pressures, many of the controller functions can be automated, at the same time increasing controller productivity. Also, new "tools"—both in the aircraft and on the ground—can enhance the efficiency of the pilot/controller team and reduce the possibility of "human errors" occurring in the ATC System. Actions underway or planned along these lines are described later.

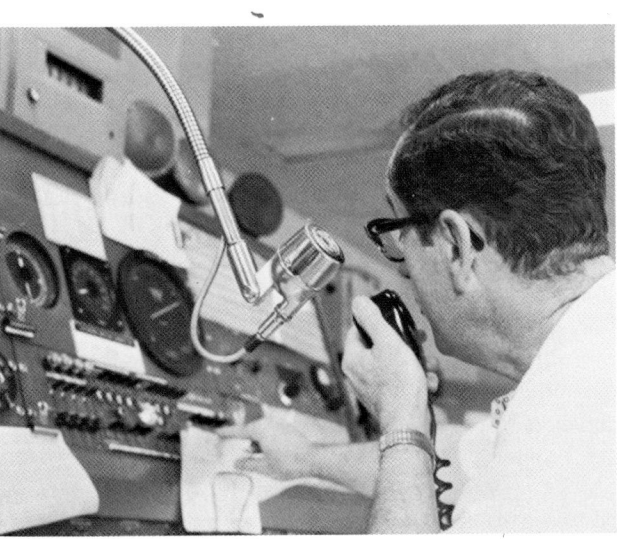

39 Patience and understanding help to smooth the way between ground and air.

CHAPTER V

40 Model of airspace structuring indicates the concept by which air traffic is controlled in varying degrees.

BASIC PROCEDURES

41 Organization and designation of airspace.

The control of air traffic follows many concepts similar to those involved in the control of surface traffic. All systems of traffic control have one basic objective—the prevention of collisions between the vehicles involved. At the same time, the systems should be designed so that they cause minimum interference with efficient traffic flow. An efficient Air Traffic Control System would be one in which the flow of air traffic is restricted only by sheer volume—not by limitations in the system itself.

Two principal methods have been employed to provide protection from the hazard of collisions between aircraft. One method is based on the concept that when aircraft are being flown in weather conditions where pilots can "see and be seen," the individual pilot is responsible directly for avoiding collisions ("see and avoid") with other aircraft, in the same manner as each individual automobile driver "looks out" for other traffic. Certain rules define the conditions under which this type of flight must be conducted, known as the visual flight rules (VFR). An aircraft flying in accordance with such rules is referred to as a "VFR flight."

The other principal method in affording protection from the hazard of midair collision relies on the ground-based Air Traffic Control (traffic management) Service. This service is designed to provide separation between aircraft operating in accordance with the instrument flight rules (IFR), primarily when weather conditions are such that pilots cannot "see and be seen," by giving instructions to the pilots as to altitudes and flight paths to be followed. An aircraft being flown under these rules is called an "IFR flight."

In addition, this type of service is provided to IFR flights when they are operated in VFR weather conditions by ensuring that the appropriate separation is maintained between all such flights. This ATC Service does not, however, provide separation between IFR flights and VFR flights under these conditions.

The fundamental purpose of the ATC System is to provide for the safe and efficient operation of aircraft. The visual flight rules and the instrument flight rules as well as their related separation standards are designed primarily for this purpose alone. In providing ATC Service, however, other factors must also be given consideration if air transportation is to function effectively. The most important factor in this category is that of expediting the flow of air traffic. Thus, it is intended that in applying the various separation standards the ATC Service also must extend every effort to prevent delays to air traffic both on the ground and in the air.

Another important factor is reliability. This means that equipment such as aids to navigation, radar, and communications must be adequate to meet user needs. All such equipment must be properly maintained so as to avoid catastrophic breakdowns which, of course, will directly affect both safety and the expeditious handling of traffic.

As pointed out previously, the Air Traffic Control Sys-

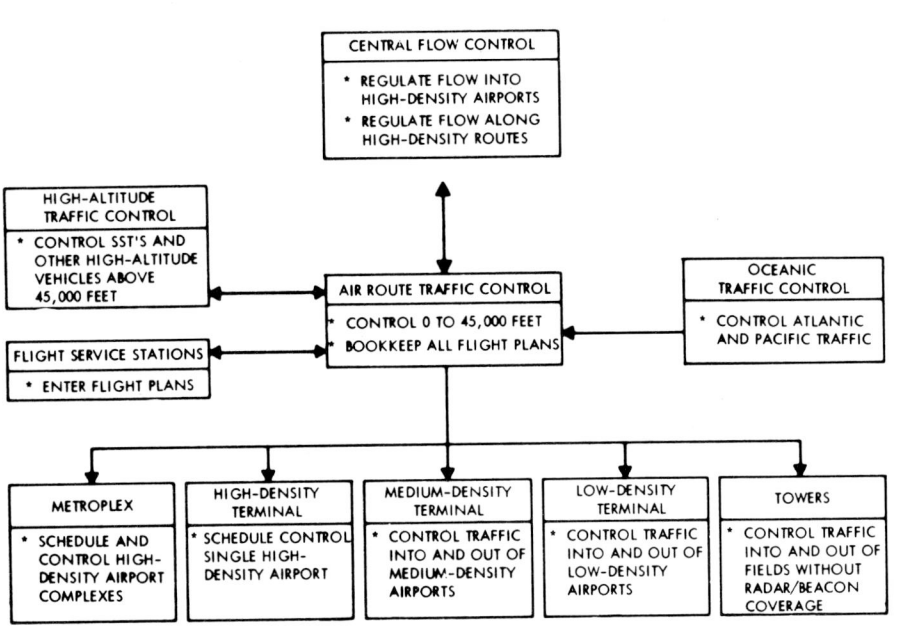

42 Structural organization of the Air Traffic Management System.

43 Control areas under the jurisdiction of air route traffic control centers in the continental United States.

tem depends upon the cooperation of both the pilot and the controller. To do their job effectively, they need certain information. The pilot needs information as to the position of his aircraft with respect to locations on the earth's surface or other navigational reference. The controller needs to know the intent, position, and identity of each aircraft in a given airspace environment.

The controller initially acquires these data by means of a "flight plan" in which the pilot provides information regarding his intended plans. The flight-plan format specified by ICAO includes the following data.

- Type of flight plan (e.g., VFR or IFR)
- Aircraft identification
- Radio identification
- Type of aircraft
- Time of departure
- Airport of initial departure
- Route to be followed
- Airport(s) of intended landing
- True airspeed
- Cruising levels
- Estimated elapsed time for each route segment
- Alternate airport(s)
- Estimated total elapsed time to airport of first intended landing
- Fuel endurance
- Radio transmitting frequencies
- Navigation and approach aids
- Total number of persons on board
- Name of pilot in command
- Identity of operator
- Emergency and survival equipment.

In the United States, this flight-plan format is considerably simplified.

The flight-plan data enter into the ATC System via any one of several methods such as a flight service station, airport control tower, airline dispatch office, or military operations office. These data are then processed in the appropriate center and surveillance of the aircraft commences on departure.

The provisions of Air Traffic Control Service throughout the world is applied in varying degrees, depending upon the need and the capability in each country to furnish such service. In some cases, a highly sophisticated national system of airways or air routes has been provided, within which a complex and highly regimented ATC System operates—such as in the United States. In other cases, a less rigid system is furnished on a localized or regional basis, and in still other cases, ATC Service is provided only at airports or other landing areas to provide limited service to arriving and departing aircraft. There are many portions of the world's airspace and many airports which have no form of ATC Service.

AREA CONTROL

Those facilities providing area or en route control service are basically the same everywhere. In general, such area control centers or air route traffic control centers, as they are termed in the United States, are normally located as a complete, separate facility, which may be located on an airport, or in other cases at remote sites. Their area of jurisdiction usually includes many thousands of square miles of airspace. These centers are provided with a complex communication network connecting them to all the airports and other aviation facilities which are within their area of jurisdiction, as well as to similar adjacent centers. In addition, they also have very complete radio communication networks with which to communicate with aircraft requiring ATC Service.

Long-range radars which electronically provide position information, are utilized to assist the controller (through surveillance) to achieve desired separation between aircraft. Some centers also have computers which automate many of the routine functions of the controller.

In order to maintain a controller's workload at a level which is within his capability to handle, the center's airspace is divided into "sectors." This airspace is a defined geographical area which encompasses a number of airways/routes, airports, and navigation aids, and is also defined vertically. Each such sector is assigned an appropriate number of controllers and assistants who are responsible for all aircraft in their designated airspace. In effect, the center's airspace is divided into small portions of the whole airspace, each of which will normally con-

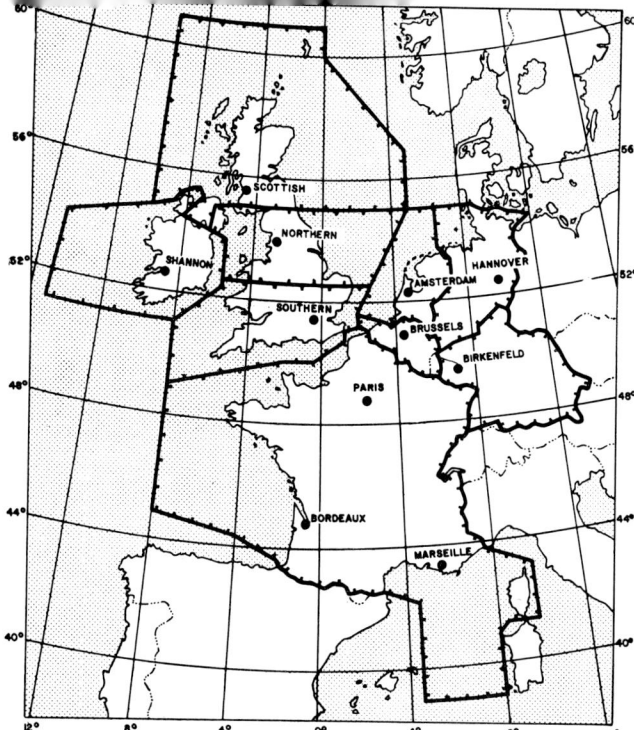

44 Western European control areas.

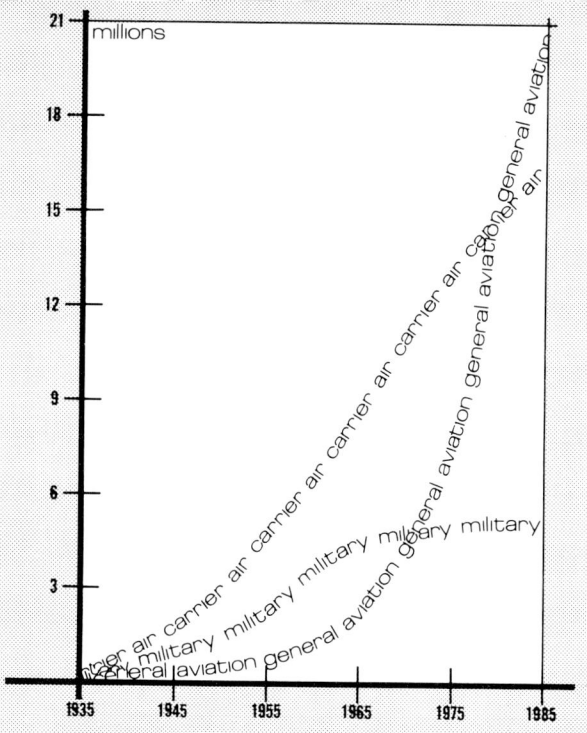

45 Distribution of IFR aircraft handled by air route traffic control centers.

tain a similar number of aircraft. Provision is made to combine sectors during periods of low traffic density and to further subdivide certain sectors when the volume of traffic reaches the point where a single controller can no longer handle the traffic. Sectors are generally classified as "manual," "radar," or "oceanic," depending upon the type of control procedures used. Each of these classes is subclassified as either "high" or "low."

Each sector has a controller who is directly responsible for the control of air traffic within his assigned airspace. The radar and communications equipment provide, in general, the means by which controllers receive position data on aircraft and through which ATC instructions are conveyed to pilots. In order to determine the correct instructions, it is essential that the controller be fully cognizant of the position and future plan of every aircraft within his sector. To accomplish this, "flight progress boards" are used on which are placed movable paper "flight progress strips" containing all the pertinent current flight data necessary to provide ATC Service. These data are derived initially from the flight plans, and each flight progress strip is divided into a number of boxes in which are inserted flight data such as the following.

- Aircraft identification
- Number of aircraft and type of aircraft
- Symbol for any special airborne equipment carried
- Filed true airspeed (plus revisions)
- Estimated groundspeed (plus revisions)
- Previous reporting point (or fix)
- Estimated time over previous reporting point
- Revisions to estimated time over previous fix
- Actual time over previous reporting point
- Time estimated by the controller when aircraft will be over current reporting point
- Arrows to show whether aircraft is arriving or departing
- Time estimated by pilot when aircraft will be over current reporting point
- Actual time over current reporting point
- Altitude in hundreds of feet
- Next reporting point
- Pilot's estimated time over next reporting point.

Additional items may be included or items may be eliminated according to the nature of the flight and traffic within a particular sector. Where a center does not have the requisite computer, flight progress strips are manually prepared and handled. When the flight progress strips are printed by a computer, relatively few items require manual printing. The strips are sequenced vertically, by time reference, and are grouped according to the reporting point to which the flight data apply.

In a sector having long-range surveillance radar (not automated), the flight data used by the controller are written on "radar target markers" or "shrimp boats" and placed on the radar display in relation to the aircraft's moving "blip." These markers are made of clear plastic in the shape of a trapezoid, with one end vertical and the other end diagonal to the parallel sides. Normally, the flight identification and altitude are handwritten by means of a grease pencil. Other data may be added, if desired by the controller. The remainder of the flight data, if required, are retained on flight progress strips adjacent to the radar display. In centers which are automated, these data are electronically superimposed directly on the radar display.

FLOW CONTROL

Flow control has been developed over a number of years as a method of "metering" the flow of air traffic into busy airports. The main objective of flow control service is to regulate or restrict the flow of IFR traffic within an affected area or at specified altitudes to the maximum number of aircraft which can be safely accommodated by the ATC System. Flow control forecasts are issued periodically to indicate the anticipated delays expected to apply during specified periods of time—usually not more than two hours. Flow control normally is applied when arrival delays will exceed 15 minutes and are expected to prevail for an extended period of time; when route segments require preventive action to avoid traffic saturation; where traffic flow is disrupted due to a breakdown in navigation facilities; or where weather conditions have caused excessive delays in executing normal landing procedures.

Advanced flow-control procedures are implemented

Separation between aircraft is established by the air traffic controller in three dimensions—laterally, longitudinally, and vertically. The effect is that each aircraft has its own collision-avoidance "box" which moves with it as it flies through the airspace. Sizes of these boxes vary according to accuracy of position-measurement facilities in the air traffic control environment encountered at any given time.

46 *Terminal Air Traffic Control System.*

during peak traffic hours as may be required to hold aircraft on the ground at points of departure until the ATC System can safely and expeditiously handle them. The appropriate center calculates the hourly demand on the affected airport or airports and then determines an acceptance rate based on forecast weather and runway configuration. When the demand is forecast to exceed the acceptance rate beyond certain tolerances, the advanced flow-control procedures are placed into effect to regulate the flow of traffic so as to distribute delays equitably among all users. This eliminates en route holding except as may be necessary when approaching the destination airport, for delays on the ground are less costly than holding in the air.

The main difference between these two forms of flow-control procedures is that the first method achieves its objective by holding aircraft in the air while en route. The advanced flow-control concept, on the other hand, holds them on the ground at the departure airport and is designed to handle the more extreme congestion problems. Various modifications to these basic systems are applied wherever congestion during peak periods presents a significant delay problem. These techniques, however, only provide temporary relief for air traffic delays. They do not present basic solutions to the underlying problem, which is to increase airport/airspace *capacity*.

ASR	— airport surveillance radar	RVR	— runway visual range
RDR BCN	— air traffic radar beacon	ARTCC	— air route traffic control center
PAR	— precision approach radar	FSS	— flight service station
ASDE	— airport surface detection equipment	TVOR	— terminal VHF omni range
ILS	— instrument landing system	TRACON	— terminal radar approach control

47 IFR room controls traffic within a radius of 25 to 60 miles from the airport at which it is located.

48 Control tower "cab" oversees surface and local airport traffic.

APPROACH/DEPARTURE CONTROL

Approach and departure control facilities, sometimes referred to as "IFR rooms" or "TRACON's" (terminal control), are located on the main airport which they serve and are usually in the tower building. When such a facility controls more than one airport, it is provided with the appropriate interphone connections to the airports under its jurisdiction. These facilities usually control an area of from 25 to 60 miles from their major airport. Arriving traffic is passed (or "handed off") from the cognizant center to the approach control sector, and the reverse process takes place for departing traffic. In certain cases, the approach control and departure control sectors may be subdivided into smaller sectors to reduce the controller workload. Approach/departure control facilities use airport surveillance radar and flight progress strips in a manner similar to that followed in centers for the control of en route traffic.

AIRPORT CONTROL

The control of air traffic on or in the vicinity of an airport is provided by an airport traffic control tower. At those locations where "approach control" service is provided, the tower normally accepts air traffic from it at about the point where the aircraft can be visually identified. When there is no "approach control" facility at the airport, its function is assumed by the tower. In most cases, towers are located on top of a high building from which the controllers are able to observe aircraft movements anywhere on the airport surface and in the surrounding airspace. The "tower cab" as it is sometimes called, is a completely enclosed glass room of varying dimensions, depending upon the number of controllers who will operate from it. The tower's "tools" include a number of radio transmitters and receivers with which to communicate with pilots and a "light gun" with which to send control signals to any aircraft not radio equipped. Some towers also have airport surface-detection radars to assist the controller in directing aircraft movements on runways and ramps.

Departing aircraft are given instructions regarding when and how they may taxi from loading ramp to the runway in use, followed by takeoff clearance when the pilot is ready and traffic permits. Arriving aircraft are handled by the control tower in a similar manner, by "clearing" the aircraft to land when airborne and ground traffic permits, and then by issuing appropriate taxi instructions to guide the aircraft to its unloading point.

At some busy airports an Automatic Terminal Information Service (ATIS) is available to pilots of departing and arriving aircraft. This service consists of a continuous radio broadcast on a special frequency of recorded and periodically updated noncontrol information, which is designed to relieve control-frequency congestion and controller workload. Included is information regarding ceiling, visibility, wind direction and speed, altimeter settings, and runway in use. Where ATIS is not available, the tower provides this information.

ALERTING SERVICE

When the pilot of an aircraft reports that his aircraft is in a "state of emergency" or when an aircraft is overdue at its destination, any of the ATC facilities may be used as a central point for collecting all pertinent information. The cognizant ATC facility forwards all such information to the appropriate rescue coordination center, in accordance with agreed procedures. The subsequent action to be taken is dependent upon the nature of the emergency. The ATC facility receiving the information utilizes all of its communications and other facilities to the extent necessary to assist the rescue coordination center.

An emergency locator transmitter (ELT) provides an important assist to the ATC System in facilitating search and rescue service for downed aircraft. These transmitters, also called aircraft locater beacons, transmit a signal on a standard emergency channel—VHF for civil aircraft, UHF for military.

In the event of a crash, the transmitter is automatically activated (or manually in the case of hand-held devices). Search aircraft then use the emitted signal to establish the position of the downed aircraft.

RESTRICTIONS

In actual practice, the ATC System functions to a certain extent as an arbiter. If several users want to fly in a given portion of airspace simultaneously, ATC makes a decision and one or more of the users might be penalized by having his flight path adjusted to the extent necessary to assure collision avoidance. The system, obviously, is unable to allow absolutely free transit of the airspace, due to the varying demands made on it by each of the different categories of airspace users.

Flight within the ATC System, of necessity, imposes certain operational and equipment requirements on the users. This involves adherence to established rules and regulations designed to minimize the possibility of mid-air collisions within the different airspace environments. It follows, in turn, that airborne equipment and pilot requirements will vary in accordance with the particular operating conditions of the different airspace users.

A major challenge, therefore, in designing and operating the ATC System—present and future—is that the *price of admission* to the system should be reasonable and *cost effective* to the user in relation to his needs. Since a flight of any user falls into either of two basic operational classifications—VFR and IFR—these rules play a very important part in the functioning of the system, and in the respectively required pilot qualifications and aircraft capability.

VISUAL FLIGHT RULES

The basis for the visual flight rules (VFR) has been developed by the International Civil Aviation Organization for international application and termed by ICAO "visual meteorological conditions." Although these rules generally are uniform in the various ICAO member countries, there are some relatively minor differences. Flights performed in accordance with VFR encounter minimum restrictions in the ATC System and are required merely to follow general rules of the road. Some restrictions—in terms of equipment—may be applicable, however, when a VFR flight operates in high-density traffic environments. The basic responsibility for avoiding a midair collision when flying under VFR rests with the pilot. ATC occasionally may give the pilot relevant traffic information derived from radar surveillance ("traffic advisories") as an assist in looking out for other aircraft.

The complete visual flight rules are quite complicated, partly due to incorporation of a variety of exceptions to the basic rules and partly to the different types of airspace within which various combinations of the rules apply. Essentially, an aircraft may **not** be flown under VFR:

- Within certain prescribed proximities of a cloud formation vertically (above and below) and horizontally.
- Unless flight and ground visibilities are equal to or better than defined distances.
- Unless a "special" VFR authorization is given by ATC if weather is below prescribed VFR minimums.

VFR Separation Standards

As discussed previously, the pilot when flying in accordance with the visual flight rules has the direct responsibility for providing separation from other aircraft by means of "see and avoid" tactics. To facilitate the ability of the pilot in exercising this capability, certain basic visibility minima for VFR flight have been established in the United States in relation to altitude above mean sea level (MSL).

Altitude	Uncontrolled airspace*	Controlled airspace
1,200 feet or less above the surface, regardless of MSL altitude	1 mile	3 miles
More than 1,200 feet above the surface, but less than 10,000 feet MSL	1 mile	3 miles
More than 1,200 feet above the surface and at or above 10,000 feet MSL	5 miles	5 miles

* Uncontrolled airspace is that portion of the airspace that has not been designated as continental control area, control area, control zone, terminal control area, or transition area, and within which ATC has neither the authority nor the responsibility for exercising control over air traffic.

The principal VFR procedural separation standard is the observance by the pilot of cruising altitudes in accordance with certain criteria. These criteria apply to flights, both in controlled and uncontrolled airspace, performed more than 3,000 feet above the surface. (Below 18,000 feet above MSL, the term "altitude" is applied, and above 18,000 feet "flight level"—FL—is used.) MSL altitudes must be appropriate to the magnetic course being flown

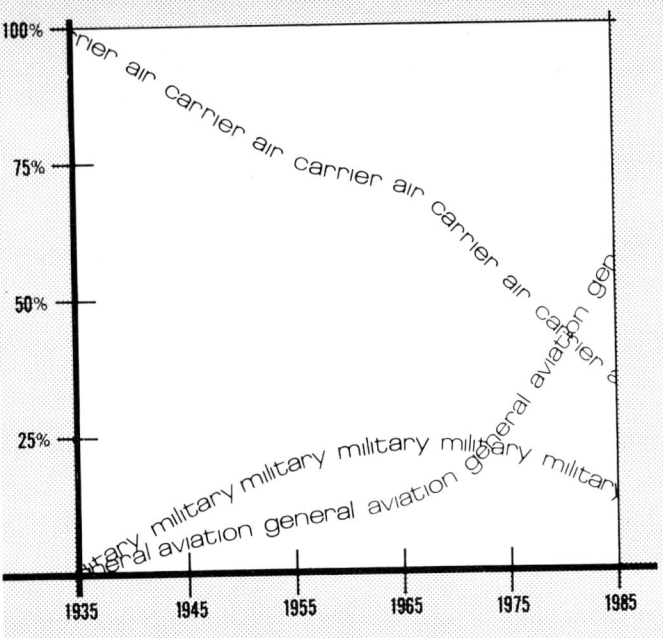

49 *Percentage distribution of instrument approaches by class of airspace user. Growing volume of general aviation* IFR *activity points to a need for the* ATC *System's orientation toward general aviation* IFR *requirements.*

following the same system as for IFR flights, except that 500-foot intervals are used below FL290 (e.g., 4,500 feet, 5,500 feet and FL185, FL195); and above FL290 4,000-foot intervals begin at FL300 and FL320 (e.g., FL340 and FL360).

If a VFR flight is performed within a designated positive control area (PCA), the aircraft/pilot requirements are essentially the same as for IFR operations, and such VFR flights are provided separation by ATC in accordance with IFR separation standards.

INSTRUMENT FLIGHT RULES

A flight that is not performed in accordance with the visual flight rules or in uncontrolled airspace, must be performed in accordance with the instrument flight rules (IFR). Basically, for an IFR flight the aircraft must be fitted with specified flight instruments, communication and navigation equipment, and the pilot must meet certain proficiency standards for instrument flying. In certain designated airspace, the aircraft must be radio equipped to provide direct pilot/controller communication on frequencies specified by ATC, as well as have an airborne radar transponder to tie into the ground radar surveillance system.

For a flight conducted under IFR in controlled airspace, the pilot must do the following.

- Submit a flight plan and thereafter comply with ATC "clearances" or traffic control instructions. (An ATC clearance is defined as an authorization for the flight to proceed under conditions specified by an air traffic control facility).
- Make periodic position reports as to time and altitude of passing specified "reporting points" (unless under active radar surveillance).
- Obtain prior approval from ATC for any desired change in flight plan or in current air traffic control instructions.

Separation Standards: General

Aircraft flying under IFR are provided separation by ATC for collision prevention purposes in accordance with two types of standards, procedural and radar. In either instance, separation is applied in three dimensions—lateral (width), longitudinal (length), and vertical (height).

When using procedural standards, the controller applies established criteria for traffic spacing upon departure of an aircraft. Thereafter, based on inflight position reports received from the pilot, the controller adjusts the flight path as necessary to maintain desired separation by instructions to the pilot, and monitors results by checking subsequent position reports.

Radar separation standards are based on direct surveillance by the controller of a dynamic traffic environment as shown on a radar display. (These standards are discussed in detail in Chapter 6.) Procedural separation standards are more wasteful of airspace than radar standards, and are used only when radar surveillance is not available.

Separation Standards: Procedural

Procedural separation standards are quite complex, and their application involves consideration by the controller of numerous factors in determining the precise standards to be applied in a given situation. These standards, however, may be summarized along the following lines.

- Lateral separation is applied so that the distance between aircraft—right or left—is never less than an established distance, taking into account the accuracies (or inaccuracies) of particular navigational systems, including a "buffer" for margin of error. Lateral separation of aircraft at the same level is obtained by requiring operation on different routes or in different geographical locations as determined by visual observation or by use of navigational aids.

- Longitudinal separation—fore and aft—is applied so that the spacing between the aircraft is never less than a prescribed minimum. Longitudinal separation is established by requiring aircraft to depart at a specified time, to lose time to arrive over a geographical location at a specified time, or to hold over a geographical location until a specified time. Time-spacing after departure is based on the pilot's and ATC's estimates of the aircraft's arrival times over the specified reporting points. The magnitude of this time-spacing depends on many factors, navigation accuracy being one, as in the case of lateral separation. Procedural longitudinal separation can be the most waste-

CHAPTER VI

RADAR

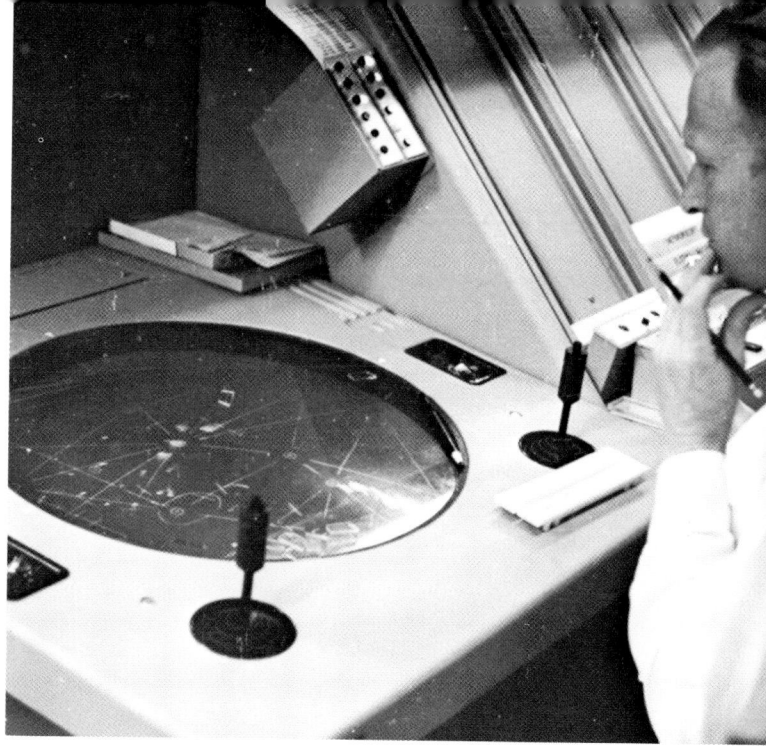

50 *Introduction of radar ushered in the second generation* ATC *System.*

51 *Radar acquisition and controller display.*

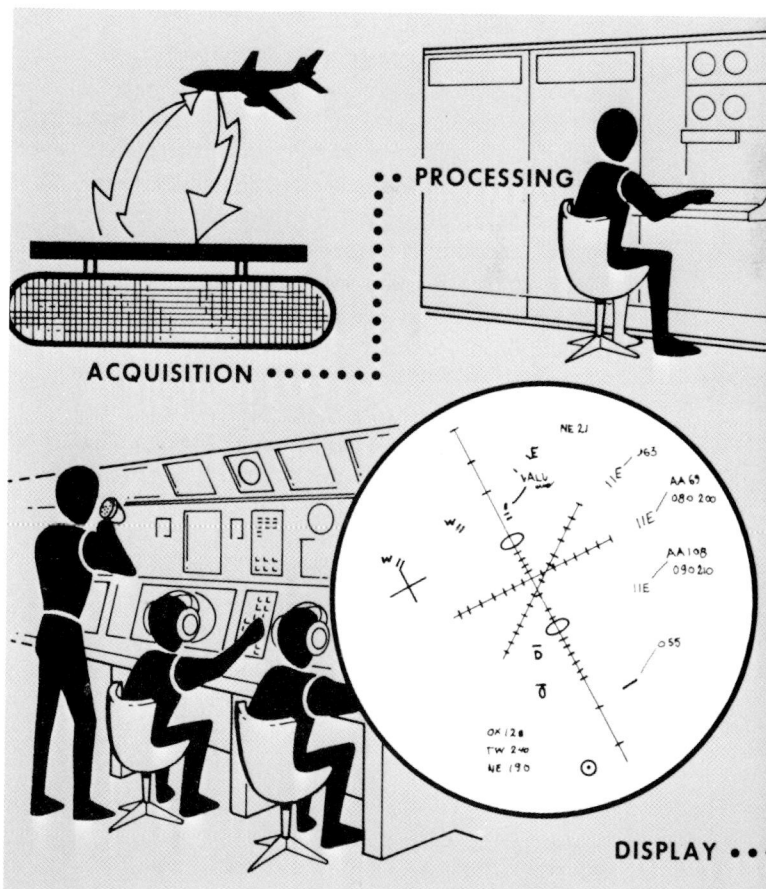

ful of airspace as, for example, a 15-minute, time-spacing interval between two 600 MPH jet aircraft means that they are separated longitudinally by ±150 miles.

• Vertical separation—height—generally is applied by the controller when satisfactory aircraft separation is not available through lateral or longitudinal separation methods. It is obtained by requiring aircraft to fly at different heights expressed in terms of "flight levels" or "altitudes." In the United States, "altitude" is used below 18,000 feet above mean sea level (MSL), and the aircraft's altimeter is set to the barometric pressure of the nearest ground station so that it indicates the actual height of the aircraft above mean sea level. Above 18,000 feet MSL, the aircraft's altimeter is set at a standard pressure reading of 29.92 inches (or 1013.2mb) and the height of the aircraft is referred to as its "flight level" (a flight level of 29,000 feet is referred to as FL290). When vertical separation is required between aircraft, 1,000 feet spacing normally is used below FL290 and 2,000 feet above FL290. Altitudes and flight levels are at odd 1,000-foot intervals (e.g., 15,000 feet, FL210) when flying magnetic courses from 0° to 179°, and even (e.g., 16,000 feet, FL220) when flying magnetic courses from 180° to 359°. At and above FL290, 4,000-foot intervals are applied beginning at FL290 when flying a magnetic course from 0° to 179° (e.g., FL290, FL330), and beginning at FL310 when flying a magnetic course from 180° to 359° (e.g., FL310, FL350).

The advent of radar (RADIO DETECTION AND RANGING) has been accepted generally as being the most important single step forward in the development of the Air Traffic Control System throughout the world since the end of World War II. Its introduction in the United States Air Traffic Control System in the early 1950s marked the beginning of the transition from the first to the second generation system.

Radar forms the most universal basis on which the Air Traffic Control Systems of the world function, and is the foundation on which programs for automation in the control of air traffic is largely based. In view of the past, present, and continuing importance of radar in the ATC System, it is useful to consider some of the broad principles of radar and how it operates, in order to understand and evaluate the role of radar in meeting problems in controlling future air traffic.

The principles of radar actually are not new; some early experiments were conducted in the 1880s by physicist Heinrich Hertz. In fact, in 1904, a patent was issued to a German engineer on a "radio-echo collision prevention device." In 1922, the famed electronics inventor, Marconi, propounded some principles which were basically those used in radar today, when he conceived sending radio signals between ships which "would be reflected back to a receiver screened from the local transmitter on the sending ship, and thereby immediately reveal the presence and bearing of the other ship in fog

52 *Radar facility in Germany.*

53 *Radar site in Norway.*

or thick weather."

Further improvements in this concept were developed, including the introduction of the "pulse" principle on which modern radar is based. Finally, successful pulse radar systems were produced independently and nearly at the same time in the United States, England, France, and Germany during the years 1935-1940. World War II brought many improvements in the techniques and equipments used in radar, especially for application with aircraft, and the stage was set for its introduction into civil aviation and air traffic control in the post-World War II years.

The application of radar in the Air Traffic Control System has involved two basic design concepts. The initial type of radar, called primary radar, began to be used in most parts of the world in the early 1950s. Another form of radar, called secondary surveillance radar (SSR), found some limited application during the early 1960s. Following subsequent agreements on international radar standards by the International Civil Aviation Organization, SSR has become the standard radar system to be implemented throughout the world for advanced air traffic control application. (When the word "radar" is used alone, it includes both primary and secondary modes.)

PRIMARY RADAR
Basic Principles

In primary radar, a beam of individual pulses of energy is transmitted from the ground equipment at the rate of approximately 1,200 per second, while the transmitting antenna rotates at a speed of 3 to 6 RPM for long-range systems, and as fast as 60 RPM for airport coverage. These pulses hit the aircraft from 16 to 34 times each scan, depending upon the rotation rate of the antenna and the width of the beam. An aircraft in the path of this radar beam will reflect some of the pulses back to the ground where they are picked up by a receiving element on the same rotating antenna. The strength of the reflected energy depends on the aircraft's size and attitude in addition to the transmitter power. This reflected energy produces a bright "echo" or "target" on a cathode ray tube (CRT), usually referred to as the "radarscope" of "PPI" (plan-position indicator). The direction of the target from the radar site is known from the direction to which the antenna is pointing when the target is received. Distance of the target from the radar site is determined by the time it takes for the radar pulse to travel from the radar site to the aircraft and back (about 1 mile in 10.75 millionths of a second). This is known as a "radar mile."

Types

The most common type of primary radar is the airport surveillance radar (ASR) which was designed as a medium-range radar—about 50 miles—for the control of traffic in the vicinity of an airport. Moving targets, fixed echoes, and areas of precipitation are displayed. The normal rotation of an ASR is approximately 13 RPM, so the traffic situation is updated every few seconds. The accuracy of this radar system is such that it can be used to assist pilots in making instrument approaches.

Another type of primary radar used in the control of air traffic is the long-range radar known as the air-route surveillance radar (ARSR). This system has a range up to about 200 miles and will detect aircraft up to an altitude of about 40,000 feet. It is used in air-route traffic control centers (or area control centers) for the control of en route traffic. The ARSR normally is provided with features similar to the ASR. Because of its slower rotation—3 to 6 RPM and other factors, its accuracy and resolution are not as high as the ASR.

A third type of primary radar equipment used less frequently by ATC is the airport surface-detection equipment (ASDE). This equipment is designed to be used in airport control towers to observe the movement of aircraft and vehicles on the surface of the airport. ASDE normally is used with a range of one mile, although its range can be as high as four miles. The accuracy and resolution of this equipment is extremely high. Runways, taxiways, buildings, as well as aircraft, can be readily identified by their physical features alone; however, many "blind spots" occur in which no targets can be received.

Another primary radar, not generally installed at civil airports, is the precision-approach radar (PAR) which is used as a landing aid (in military use this system is known as ground controlled approach or GCA). PAR has a normal range of about ten miles, but its pulses are transmitted only on the final approach to a runway. A PAR uses two antennas, one scanning a vertical plane, with an elevation of 7° and the other scanning horizontally, giving an azimuth of 20°. The controller's console is provided with two displays, one having a range of ten miles and showing both azimuth and elevation, with the other having a range of three miles and the same type of data is displayed. PAR permits the controller to aid the pilot in making satisfactory approaches in extremely poor weather. It may also monitor instrument approaches made by other systems, e.g., ILS. Its high degree of accuracy permits the detection of 300 feet in range and variations of 10 feet in elevation and 20 feet in azimuth at a distance of one mile.

What the Controller Sees on Primary Radar Displays

On his PPI, the controller sees radar targets as "blips" representing aircraft in the range of his primary radar, moving at various speeds and in various directions. The radarscope also usually displays a video map of the controller's assigned airspace so the controller can determine the position of the aircraft targets relative to such features as navigational aids, reporting points, airports, and airways. There is nothing, however, to tell the controller which airplane corresponds to which target. Since the entire vertical section of the airspace scanned by the radar is shown on the scope, targets will appear to merge when, in fact, they may be separated vertically by many thousands of feet in altitude. To determine the altitude of the identified targets, the controller must rely on voice reports from pilots.

Before separation between aircraft can be effected by reference to the radar targets, aircraft involved must be identified positively prior to any action by a controller. Aircraft may be considered to be identified when at least one of the following criteria applies.

- A departing aircraft is observed on radar within one mile of the end of the runway.
- A position report is received from an aircraft, based on ground navigation aids, coupled with a heading consistent with the observed radar track.
- Turns of at least 30° identify a particular aircraft from all other observed targets.
- An aircraft target is observed along a line of position utilizing ground-direction finding equipment coupled with a reported heading consistent with the observed track.

Due to the possibility of an incorrect identification, the use of only one of the above methods is considered as a minimum. Whenever any doubt exists, the controller uses at least two of these criteria.

In order to maintain the correct association between the aircraft target and the flight data, while the targets are moving over the scope, the controller must provide a suitable "link." When a vertical or slanted PPI is used, the controller indicates the identification of the target by means of a "grease" pencil, with the data written on the face of the scope. In the case of horizontal PPI displays, radar target markers generally are used as described earlier.

Ground obstructions and elevations in terrain, as well as aircraft, will appear on a basic PPI display. Since the ground targets usually return the strongest echoes, they frequently mask the aircraft targets which the controller needs. (These unwanted echoes are called "clutter.") To overcome this problem, a technique known as a moving target indicator (MTI) was developed and is provided in most radar systems. By showing on the PPI only those radar reflections which are moving, most of the clutter is removed from the display, thus enabling the controller to see the aircraft targets more distinctly.

Another similar problem arises by echoes received from precipitation, when aircraft targets are frequently masked by the precipitation clutter. MTI is effective with this problem only to a limited extent since it permits any echo which is moving to be displayed. A technique known as circular polarization (CP) prevents the display of most of the clutter caused by precipitation. In most primary radar equipment, the controller is provided with

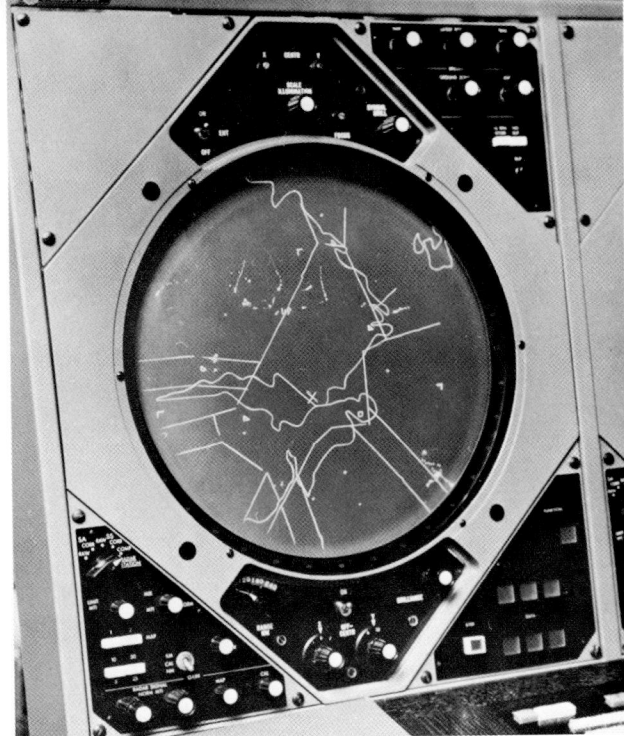

54 *Electronic daylight display used in Sweden's* ATC *facilities.*

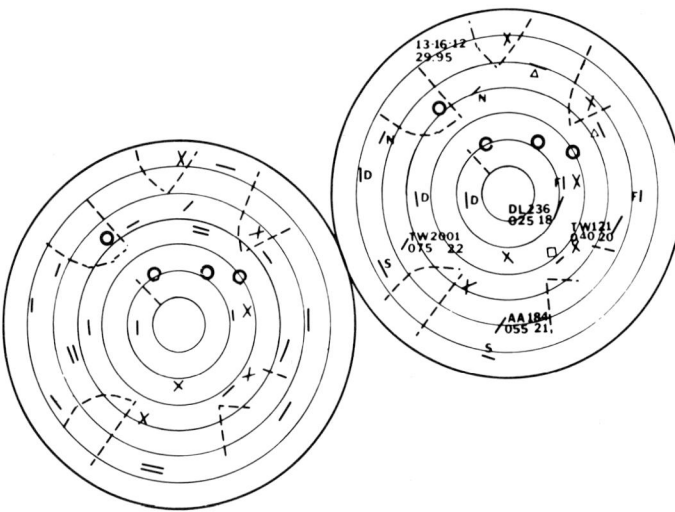

55 *Primary radar targets before and after conversion to alphanumeric* SSR *display.*

the capability of using MTI alone, MTI plus CP, or CP alone, thus allowing him a considerable amount of flexibility in his display.

Primary radar displays must be viewed in comparative darkness because of the low visual level of the displayed data. This problem requires that control rooms be operated at a reduced lighting level when the actual control of air traffic by means of radar is taking place. The use of primary radar in control towers was virtually discontinued at one time, because of the impossibility of darkening the tower cab, although it was sometimes used at night or when the ambient light in the tower was reduced due to fog.

However, equipment has been developed and is installed in many ATC facilities around the world, which, in effect, transfers the radar "picture" from the PPI to a TV display. This equipment, known as scan conversion, permits controllers to utilize primary radar under normal room lighting conditions, thus reducing to a considerable extent eyestrain and the strenuous concentration previously required. A further development of this technique is a display which can be used in a control tower even in bright sunlight.

SECONDARY SURVEILLANCE RADAR

A refinement which eliminates, to a large extent, a number of the problems applicable to primary radar is secondary surveillance radar (SSR). While generally co-located with a primary radar, this basically is a completely separate system which is capable of independent operation. In the United States and some other countries, SSR is referred to as the ATC radar beacon system (ATCRBS). The SSR system comprises a ground interrogator transmitter/receiver, antenna system (which may be associated with the primary radar antenna system), a transponder in the aircraft, and a display on a radarscope.

Basic Principles

SSR is intended to provide the air traffic controller with continuous, reliable, and accurate information concerning the plan-position (rho–theta), altitude, and identity of any or all transponder-equipped aircraft in the airspace under his jurisdiction.

Modes: The SSR system provides for six modes and their associated functions as follows:.
- Mode 1—Military.
- Mode 2—Military.
- Mode 3/A—To initiate transponder response for identification and tracking.
- Mode B—In some parts of the world during a transition period, to initiate transponder response for identification and tracking.
- Mode C—To initiate transponder responses for automatic pressure altitude transmission.
- Mode D—For future expansion of the system to study ICAO operational requirements. No functional need has been defined.

It should be noted that there are no plans for the use of modes B and D in the United States.

Identification Coding: The SSR is a valuable tool for automatically identifying aircraft, as well as for providing radar target reinforcement. Identification is achieved by providing the controller with the specific radar beacon target identity of equipped aircraft. A total of either 64 codes or 4096 discrete reply codes are available, with an additional pulse made available for "special position identification" to be transmitted on request of a controller. Two codes are reserved for transmission of a distinct emergency, code 7700; and radio communication failure, code 7600.

Altitude Transmission: SSR also provides for automatic aircraft altitude transmission in 100-foot increments. This altitude transmission capability is used to accomplish the following.
- Reduce the volume of communications between controllers and pilots by obviating the need for oral altitude reports.
- Improve utilization of airspace to climbing and descending aircraft.
- Enable the controller to be assured that vertical separation between two aircraft is being maintained.
- Enable the controller to establish greater than normal

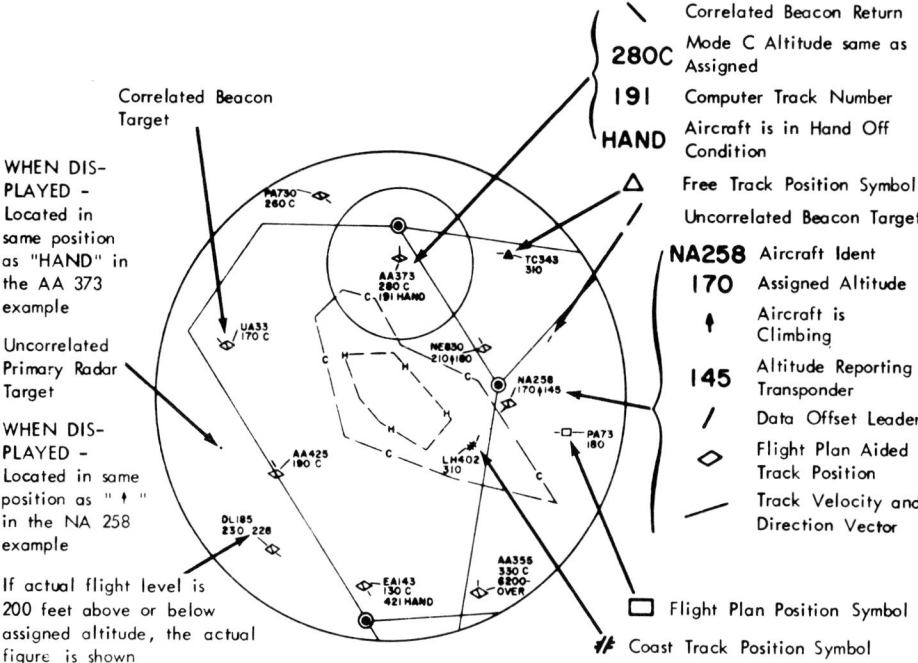

56 · *Examples of symbology used in SSR radar display.*

vertical separation in special situations such as turbulence.

SSR *System Description:* In operation, an interrogation pulse-group transmitted from the interrogator-transmitter unit, via the antenna assembly, triggers each airborne transponder located in the direction of the main beam, causing a multiple-pulse-reply group to be transmitted from each transponder. These replies are received by the ground receiver and, after processing, are displayed to the controller. Measurement of the round-trip transit time determines the range (rho) to the replying aircraft while the mean direction of the main beam of the interrogator antenna, during the reply, determines the azimuth (theta). The arrangement of the multiple-pulse reply provides individualized altitude and identity information pertaining to the responding aircraft.

What the Controller Sees on SSR Displays

With secondary surveillance radar, the controller sees aircraft returns on his PPI as two slashes, clearly distinguishing them from primary targets which are single blips. The pilot can, at the controller's request, push the "ident" button on his transponder control panel which causes the space between the slashes on the PPI to be filled in, thus forming a single wide slash. This serves as a radar identification of the target, eliminating the aircraft maneuvering usually required when primary radar alone is used.

With the 4069 code airborne transponder equipment, each flight can be assigned an individual identification code by ATC. Altitude information can also be transmitted if the aircraft is suitably provided with Mode C. When the ATC facility is equipped with a computer and more sophisticated PPI equipment, the aircraft identification and altitude data are processed and displayed in "data blocks" along with ground speed in a format of letters (alpha) and numbers (numerics). Each data block is connected to the associated radar target by a "leader." The displays can be adjusted so that only those aircraft at selected altitude levels will appear on a particular display, thus limiting a controller's area of surveillance to those aircraft which are of most immediate concern to each other. (This "alphanumeric" ATC display system is described in more detail in Chapter 7.)

RADAR CONTROL

The use of radar for air traffic control provides the controller with a current, instantaneous indication of the positions of all aircraft under his jurisdiction, whether he is handling en route or terminal air traffic. Some of the major uses of radar in the ATC system are to expedite arrivals and departures; resolve en route conflicts; and vector aircraft to eliminate congestion.

An aircraft may be "vectored" by a controller to provide desired lateral or longitudinal separation between aircraft or to meet noise abatement requirements. Vectoring is accomplished by instructing a pilot to make a turn in a specified direction and to a specified magnetic heading. The controller also may give the pilot "speed control" instructions in order to achieve optimum longitudinal separation between aircraft in-trail.

In general, the more extensive use of radar in the control of air traffic considerably decreases the amount of airspace assigned to each aircraft in comparison with that needed when the earlier described procedural system is used. The use of radar also permits more expeditious handling of aircraft, particularly in the terminal area.

A pilot of a VFR aircraft may request an air traffic controller to provide him with radar information regarding the position of other aircraft in his immediate vicinity. This service is referred to in the United States as "VFR radar advisories" and may be provided if the controller's workload permits.

One of the problems in using radar for air traffic control purposes has been to relate the positions of aircraft on the controller's PPI's to weather conditions actually being encountered by the pilot. A controller might vector an aircraft for traffic separation purposes in a way that would actually lead a pilot into severe turbulence or hazardous meteorological conditions. One approach to solving this problem has been to integrate current weather contours on the controller's PPI to show the actual weather cross-section in the area covered by the display. The problem is more susceptible to solution in the high-altitude radar presentations than those used for low al-

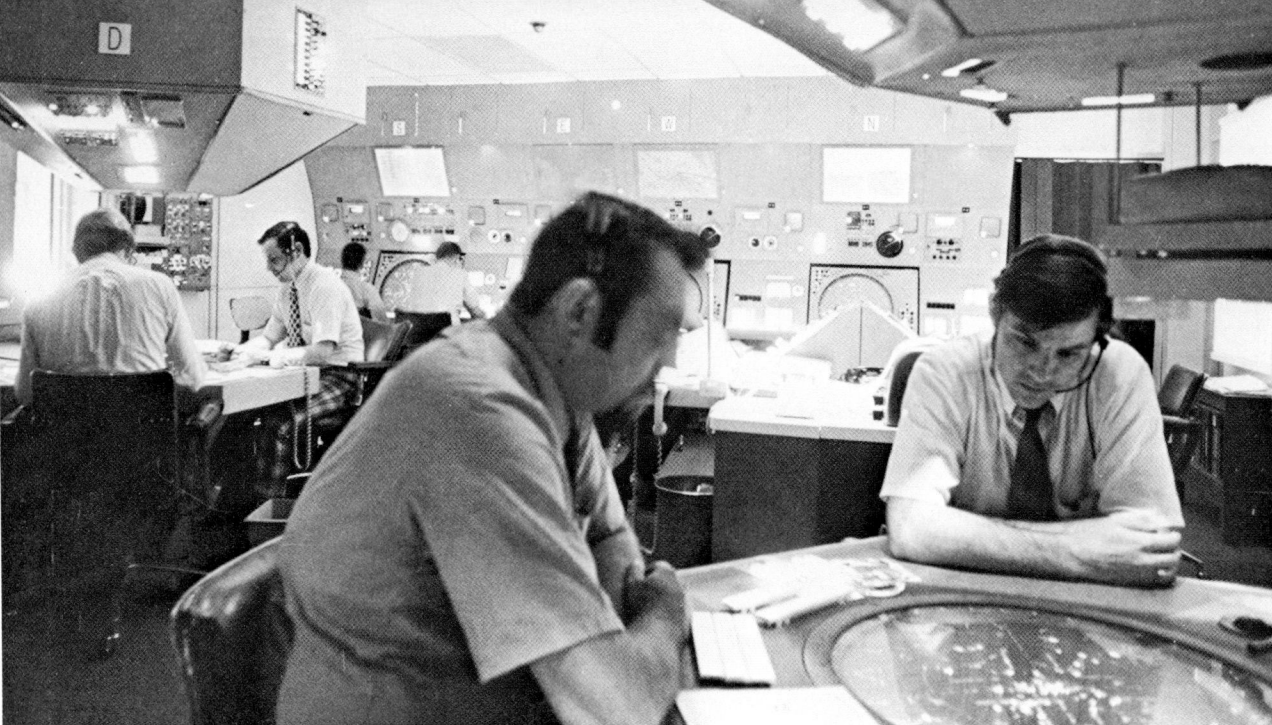

57 *Continuous vigil of radarscopes is needed to assure that the required radar separation standards are applied at all times.*

titudes. Another approach is to digitize the weather information for incorporation on the controller's computer-processed PPI display. In the case of aircraft having airborne-radar weather-detection equipment, this equipment provides a backup to supplement the ground-radar weather-detection capability.

RADAR SEPARATION STANDARDS

Since radar surveillance provides the basis for the control of air traffic in most areas of the world having medium- to high-density air traffic volume, it is obvious that "radar separation standards" play a key role in determining the capacity and efficiency of the ATC System. Radar separation standards are quite complex, but a thorough analysis of these standards is needed in order to set the stage for understanding other aspects of the ATC System, both in current and future time frames. Although in general there is similarity in these standards as between the various national systems, there nevertheless are differences resulting from such factors as varying radar accuracies and coverage. The United States radar separation standards are amongst the most comprehensive and highly developed anywhere in the world, and consequently these are analyzed in detail.

General

In general, radar separation may be applied between the following.

- Radar-identified aircraft.
- An aircraft taking off and another radar-identified IFR aircraft, when the aircraft taking off will be radar identified within one mile of the runway end.
- A radar-identified aircraft and an IFR aircraft not radar-identified, when the former is climbing or descending through the altitude of the latter and the following conditions exist.

   The performance of the primary target is being displayed on the PPI being used.

   The airspace in which separation is applied is not less than 6 miles (10 miles, if 40 miles or more from the radar antenna) from the edge of the radar PPI display.

   Flight data on the IFR aircraft not radar-identified indicates it is a type which can be expected to give adequate primary return in the area where separation is applied.

   The radar-identified aircraft is vectored on a flight path different from the route of the IFR aircraft not radar-identified before descent or climb.

   Radar separation is maintained from all observed primary and beacon targets until nonradar separation is established from IFR aircraft not radar-identified.

Target Separation

Radar separation may be applied (referring to PPI display):

- Between the centers of primary radar targets (blips).
- Between the ends of SSR control slashes.
- Between the end of an SSR control slash and the center of a primary target.

SSR Range Accuracy

SSR targets may be used for separation purposes if SSR range accuracy is verified by one of the following methods.

- SSR and primary targets of the same aircraft (not necessarily the one being provided separation) are correlated to assure that they coincide.
- When SSR and primary targets of the same aircraft do not coincide, they are correlated to assure that any SSR displacement agrees with the specified distance and direction for that particular radar system.

If SSR range accuracy cannot be verified, SSR targets may be used only for traffic information.

RADAR SEPARATION MINIMA

General

Aircraft are separated by the following minima.
- If less than 40 miles from the radar antenna — 3 miles.
- If 40 miles or more from the radar antenna — 5 miles.

Passing or Diverging

Vertical separation between passing or diverging aircraft may be discontinued when the following conditions are met.

58 Radar-separation "boxes" are established by the controller in relation to radar accuracy measurement in particular environments.

- They have passed each other and their primary targets or SSR control slashes do not touch.
- Their courses diverge by at least 15°.

Adjacent Airspace

If coordination between controllers in adjacent sectors has not been effected, radar-controlled aircraft are separated from the boundary of adjacent airspace in which radar separation also is being used, by the following minima.

- When less than 40 miles from the radar antenna — 1½ miles.
- When 40 miles or more from the radar antenna — 2½ miles.

Radar-controlled aircraft are separated from the boundary of airspace in which nonradar separation is being used by the following minima.

- If at a constant altitude when less than 40 miles from the radar antenna — 3 miles; when 40 miles or more from the radar antenna — 5 miles.
- If climbing or descending, regardless of distance from radar antenna — clear of airspace boundary.

Edge of Scope (PPI)

A radar-controlled aircraft climbing or descending through the altitude of an aircraft that has been tracked to the edge of the PPI display is separated from the edge of the scope by the following minima until nonradar (procedural) separation has been established:

- When less than 40 miles from the radar antenna — 3 miles from edge of scope.
- When 40 miles or more from the radar antenna — 5 miles from edge of scope.

Obstructions

- Within 40 miles of a radar antenna, aircraft are separated from prominent obstructions shown on the radarscope (displayed on the video map, scribed on the map overlay, or displayed as permanent echo) by a minimum of 3 miles.
- Vertical separation of an aircraft above a prominent obstruction, which is displayed as a permanent echo, may be discontinued after the aircraft has passed it.

Final Approach Course Interception

Arriving aircraft are vectored to intercept the final approach course to the end of the runway before reaching the approach gate and before intercepting the final approach glide slope. Radar vectors are applied in such a manner that interception of the final approach course does not exceed defined interception angles, depending on aircraft characteristics, and is within a prescribed minimum distance from end of runway.

Separation Boxes

The obvious effect of applying separation minima in three dimensions to aircraft is that the ATC System provides an airspace "box" for each aircraft. These boxes range in a favorable radar environment from dimensions of 1,000 feet high, 3 miles wide, and 3 miles long, to a box in a poor measurement environment with dimensions of 2,000 feet high, perhaps 120 miles wide, and upwards of 100 miles in length. There are, of course, intermediate-sized boxes used in between these extremes. It is, however, a rare occasion when any single flight can be accommodated by the optimum box size for the entire length of the flight. This is a natural consequence of a system having so many widely different measurement tolerances through which the air traffic flows.

RADAR DEFICIENCIES

Since the Air Traffic Control System places such a high reliance on radar, it is useful to review some of the deficiencies which can affect the efficiency of radar control.

These may be briefly summarized as:

- Lack of sufficient low-altitude coverage, without a very considerable increase in the number of radar stations.
- Lack of adequate returns from some aircraft not equipped with transponders.
- Unreliability, failure, or malfunction of the radar equipment.
- Blind spots in the radar pattern.
- Lack of discrimination between targets which are within 3° of each other from the radar site.
- Since radar has about a 3° accuracy, the position of an

CHAPTER VII

AUTOMATION

59 Early automation system developed in the United Kingdom for transatlantic air traffic control.

aircraft at 114 nm, for example, would have a circle of error of ± 6 nm (12 nm diameter).
- The large target "slashes" on displays can cause overlap of targets, and thus require that the controller establish excessive separation between aircraft. Radar digitizing improves this condition somewhat.
- Reduction in signal strength in certain atmospheric conditions may result in the loss of targets.
- When any two transponder-equipped aircraft are within 3.3 nm of each other in slant range and are swept simultaneously by the same interrogation beam, their reply-pulse trains will overlap and become mixed within the decoder, thus producing garbling. Depending upon the exact amount of overlap, garbling may produce:

  False target between normal targets
  False emergency alarm
  Cancellation of all or part of one or both targets (i.e., "lost" targets)
  False data readouts of identification and/or altitude, or
  False identification responses from aircraft.
- Transponder signals may be blanked out from aircraft when turning (also interference from VOR/DME).

Certain engineering and design refinements can minimize or eliminate some of these deficiencies, but most of them are basic to the inherent characteristics of radar technology. Future ATC System planning must take these problems into full consideration.

The introduction of automation in the control of air traffic, as in any field of activity, substitutes mechanical processes for certain actions otherwise performed by the human being. Examples are an automatic pilot in an aircraft, automatic posting of stock market quotations, automatic calculations of satellite orbits, or automatic tracking of space vehicles.

In its first and second generation concepts, the world's Air Traffic Control systems were essentially *manual* systems. Flight-plan computations, analyses of traffic conflicts, and presentation of flight data were performed "by hand." Communications between communicators/dispatchers/controllers/pilots were largely carried out "verbally." The control of air traffic did not catch up with modern automation technology until the early 1960s when certain automation processes were introduced which, in effect, marked the beginning of the third generation ATC System.

Benefits resulting from the conversion of the manual ATC System into an automated system are dependent upon the reliability and accuracy of the data fed into it, the efficiency of its program, and the human interpretation given to its final output. Automation speeds up and makes more efficient the processes of air traffic control; it raises the margin of safety by minimizing human-element errors; but it does not replace the controller or pilot in performing their ultimate tasks of judgment and executive decision. Automation is a major step toward updating the Air Traffic Control System. It is one of the essential ingredients for increasing the capacity of the system to meet the demands being placed upon it by the world's ever-increasing air traffic volume.

In the United States, ATC automation has been studied since the early 1940s (the author initiated an experimental data-processing installation in the Washington, D.C., Air Traffic Control Center in 1940.) In 1947, the Cornell Aeronautical Laboratory published a comprehensive report on *A Preliminary Survey of Computer Applications to Air Traffic Control.* No significant progress was made in this important aspect of air traffic control, however, until the middle 1950s, when a data processing and display system was evaluated at the FAA's Technical Development Center in Indianapolis, Indiana. In 1957, a test bed computer was utilized at the Indianapolis Air Route Traffic Control Center to process flight data and to print flight progress strips. This computer was superseded by a more advanced computer in 1959 to perform similar functions.

In 1957 the British Ministry of Aviation commenced development of a computer system to handle transatlantic air traffic. An important symposium covering ATC automation took place in Stockholm in 1960 to introduce automatic concepts as developed in Sweden. The first international conference on automation in air traffic control was called in 1961 under the auspices of the International Civil Aviation Organization. This conference re-

60 Computer room in modern air route traffic control center.

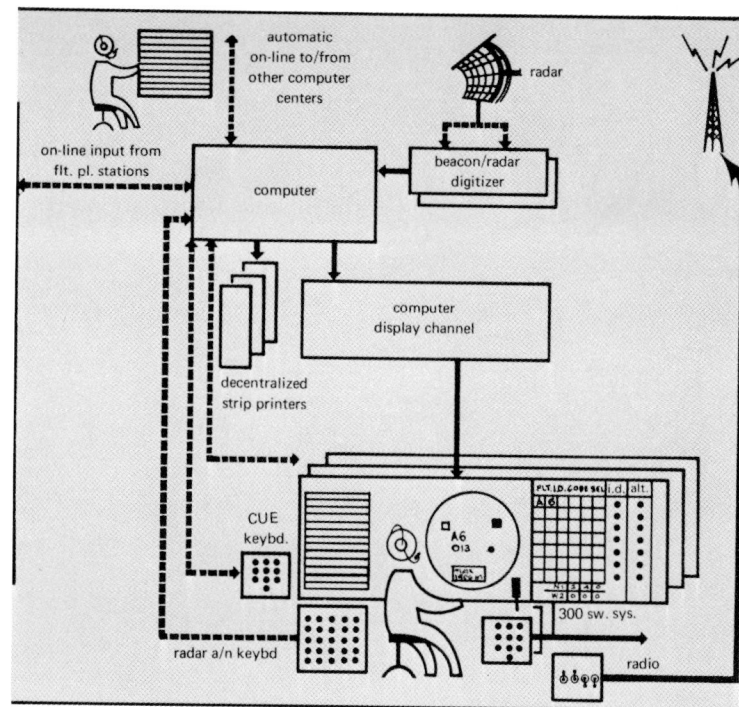

61 Functional design of "NAS Stage A" for air route traffic control centers in the United States.

sulted in the creation of an ICAO Air Traffic Control Automation Panel to coordinate development of international standards in this field. Subsequently, ATC automation studies and implementing actions are carried out on a continuing basis in many countries.

Automation in ATC covers two basic fields.
- Automatic data processing involving the automatic exchange, analysis, and presentation of flight information for controllers.
- Automatic ground/air/ground communications involving controller-pilot-controller automatic communications.

Fundamental to any automation program is the computer. Computers function in what is called a "binary mode" which simply means that the computer components can indicate only two possible states or conditions (like "on" and "off"). Basically, representing data within a computer is accomplished by assigning or associating a specific electronic signal — or absence of the signal — to signify a decimal value. This value may be expressed as an output in numbers, letters of the alphabet, or special characters. The BINARY DIGIT, or "bit," is considered as the smallest possible but still identiable measure of intelligence in communications. (Thus, "bit rate" — expressed in number of bits per second — is an index of communications speed.) An "instruction" consists of data words resulting from the transfer of bits, in variable quantities, and is a term which refers to regulating or ordering a computer to take a certain action.

AUTOMATIC DATA PROCESSING

Significant progress in the development of operational equipment has been achieved in the field of automatic data processing, the heart of which is the ATC computer system. This system must function in "real time" — i.e., no delay in the processing of the flight data —, be capable of capacity expansion, and operate twenty-four hours a day, seven days a week, without any failure which could adversely affect ATC operations. Constant self-checking for any system error must also be incorporated as a computer capability.

An example of the air traffic control central computer complex (CCC) consists of the following major elements, expressed in commonly used terms:

- Computing element (CE)
- Storage element (SE)
- Input-output control element (IOCE)
- Peripheral adapter module (PAM)
- Tape control unit (TCU).

Computers have different capacities, depending upon the manufacturer and operational requirements. Growth of a computer is accomplished by the addition of CE's — for computing power, SE's — for storage, and IOCE's — for additional input-output capability. The PAM's and TCU's in turn must keep up in capacity with the IOCE channels. The PAM provides data linkage to the system and plays a key role in automated ground/air/ground communication; the TCU provides the source for reconstruction of data — the basis for checking control efficiency, training, and accident investigation.

All elements of the CCC contain error-checking facilities to provide signals which warn of incomplete or incompatible input; or excessive temperatures, power failures, or improper operation of control circuits. Malfunction signals automatically switch elements to backup power supplies or replace failing elements with redundant elements.

Computer systems can be more versatile through incorporation of "multiprocessing" or "multichannel" modes of more than one computing element in a cooperative (parallel) relationship to perform a given job. Each element, however, can operate as a totally independent subsystem. (A multichannel ground ATC computer system can process 500,000 or more instructions per second; storage capacity can be 2,000,000 or more alphanumeric characters.)

Fail-safe and Fail-soft Concepts

Fail-safe and Fail-soft concepts have been developed for the automated ATC System to enhance its operational availability. The fail-safe concept requires that a single failure of any kind does not degrade system performance. This is achieved by providing at least one fail-safe module of every kind, which is automatically called in by the computer program when required. Likewise, multiple

62 Air route traffic control center employing NAS Stage A automation.

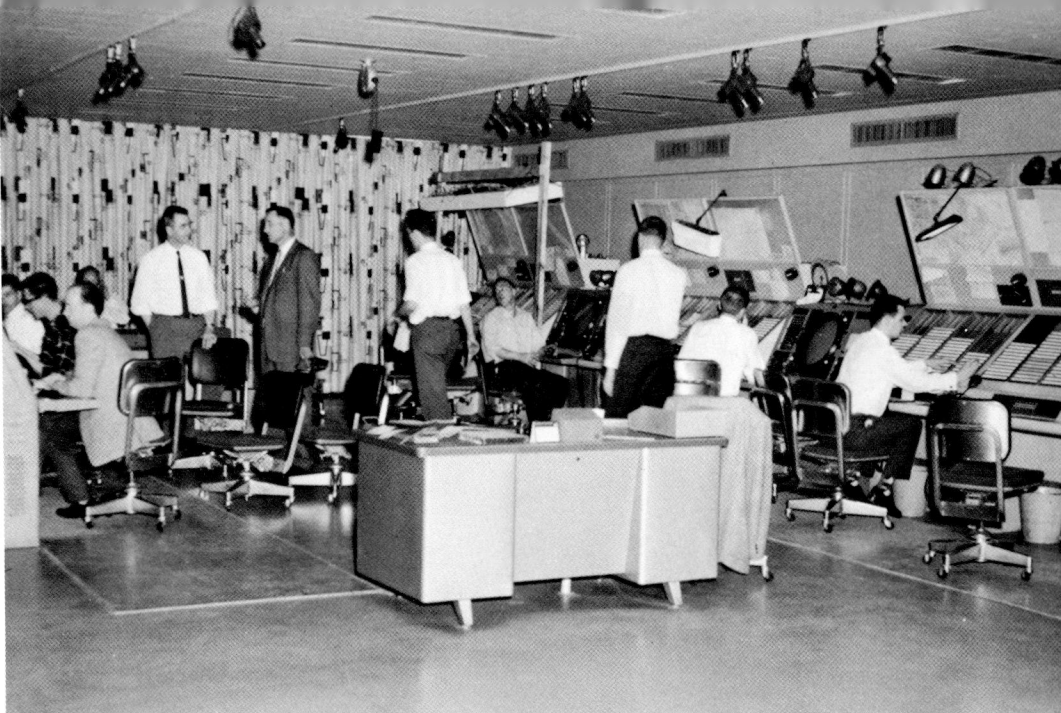

63 Early semiautomated air route traffic control center in Canada.

failures do not degrade system performance, provided that such multiple failures do not occur in modules of the same kind.

If multiple failures occur in modules of the same kind, and as long as at least one module of this kind is still in operation, the system is capable of fail-soft operation. Under fail-soft conditions, system performance is degraded. Those automated ATC systems having only two modules of a given type do not have fail-soft capabilities in the event of multiple failures of the two-element equipment group.

For those ATC systems which are provided with sufficient elements to utilize the fail-soft concept, several degrees of fail-soft capability may be available depending upon the failure status of the system.

• The frequency of operation of one or more functions is decreased according to a predetermined priority assignment.

• One or more operational functions or subfunctions are eliminated by using the same priority assignment as above.

• System capacities (e.g., the maximum number of flight plans and/or tracks which would ordinarily be stored at any one time) are reduced.

For those systems which do not contain sufficient modules of all types to utilize the entire fail-soft concept, the above items are implemented for those equipment modules which have sufficient numbers to permit fail-soft operation.

Implementation of the fail-soft concept utilizes the appropriate portions of the system monitoring and control function of the computer program to reschedule the ATC tasks and provide dynamic modification of storage allocations. The rescheduling procedure could utilize several versions of the program, stored on tape, when each version matches a different level of system functional capability to a different set of available computer modules.

In view of the inevitable reliance more and more on automation in the ATC System, effectiveness of the fail-safe/fail-soft concepts will have a direct bearing on the confidence which can be placed in overall system safety and efficiency.

Design Factors

In considering the question of automatic data processing for air traffic control on a worldwide basis, an important factor obviously is that air traffic volume, and hence data-processing requirements, varies greatly in different areas. Programming for automatic data processing takes into consideration the actual requirements of air traffic control — and the economic justifications — as they exist in different locations throughout the world.

Thus, an automatic data-processing system must be capable of being implemented on a stage-by-stage basis, starting out with the simplest automation, progressing as air traffic requirements indicate. Each step must have been a logical and useful part of an overall planned, ultimate automated system.

Automatic data-processing systems must be so designed that they are compatible, regardless of the manufacturer or the country in which the system is used. Since air transportation is international in character, and especially in view of the increasing speeds of modern aircraft and future aircraft such as the supersonic types, flight data must be readily and immediately interchangeable directly by automated means between air traffic control facilities virtually anywhere in the world. At the same time, the automatic data-processing system must be able to function in relatively small-area environments of high-density traffic.

Basic Objectives

Automatic data-processing systems for ATC embrace one or more of the following objectives.

• Flight Data Processing: This function rids the controller of certain bookkeeping and clerical tasks by printing out, automatically, flight progress strips derived from flight plans and aircraft position reports fed into the computer either by the controller or by external sources such as other centers, airline offices, or control towers. Repetitive-type flight plans, as for example those for scheduled flights, can be pre-stored in the computer and automatically processed at the appropriate time each day.

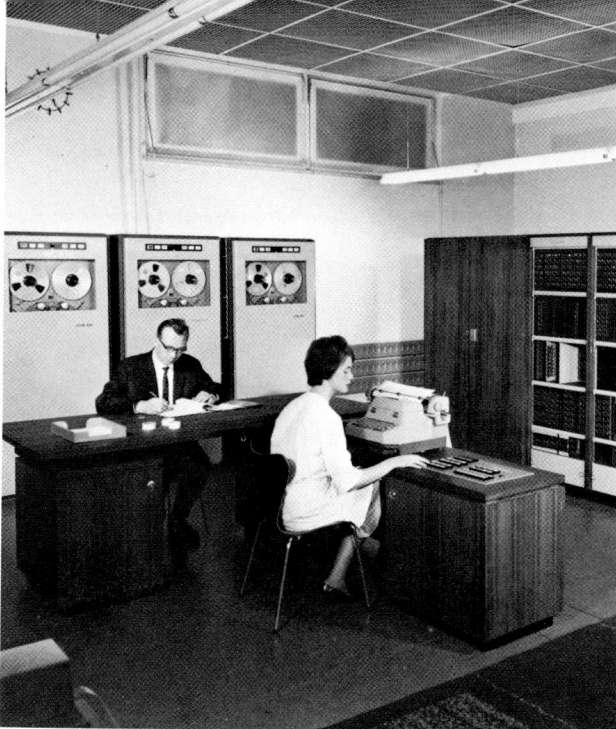
64 *Computer room in German control facility.*

65 *Area traffic control center in Sweden.*

- Flight Data Updating: This function provides for automatically updating and printing the flight progress strips and for automatic distribution of the updated strips to appropriate control sectors within an ATC facility.

- Radar Data Processing: This function can be accomplished at the same time as the initial steps in automation, or as a subsequent step. Primary radar as a data source for a computer is difficult to utilize because the radar targets do not carry information as to identity of the aircraft or its altitude (these have to be fed into the computer by the controller). Secondary surveillance radar (SSR) data are readily susceptible for feeding into the computer system. But the availability of SSR data depends on the extent to which aircraft in the traffic environment are equipped with the necessary airborne transponders; also, ATC facilities must be equipped to receive these transponder data.

- Automatic Tracking: Automatic tracking (following) of radar targets provides the basis for many functions that result from the introduction of automation into the ATC System. By tracking the targets, the computer can determine the direction and speed of the aircraft and make predictions as to potential traffic conflicts, as well as analyze and present conflict resolutions.

Flight-plan data, including the proposed route, speed, altitude, and SSR transponder code (when assigned), are used by the computer to establish the tracking function. The processing of tracks by the computer for subsequent display of primary targets and those which have not been assigned individual SSR transponder codes require their insertion into the computer manually by the controller. This process is termed "acquisition" of targets. Track acquisition of targets which have been assigned individual "discrete" codes is performed automatically by the computer.

There are two types of track initiation or reinitiation capabilities in the computer, namely, "flight-plan-aided tracking" (FLAT) and "free tracking." In the former, flight-plan-derived data are used in conjunction with radar data to initiate and maintain the track. In the free-tracking mode, controller-inserted or present track data are used to initiate the track. In both modes, provision is made to change a track to a "coast" status (either FLAT or free coast). In the "FLAT coast" status a track is maintained by "dead reckoning" using heading and speed data based on the flight plan. In the "free coast" status, a track is maintained by "dead reckoning" using controller-entered or current track and speed.

The capability of providing for the automatic tracking of radar targets is extremely valuable to the controller in that it maintains the proper correlation between the planned route and the actual route being detected by the radar. This also maintains the association between the targets and their related alphanumeric data blocks. In view of its importance to the various automated control functions, automatic tracking must have a high degree of reliability. One way — but costly — to achieve this objective is through use of multiple radars. This arrangement permits the ATC computer to select the radar which supplies the best target, and after this evaluation has been accomplished, the most reliable data are displayed to the controller.

THEORY OF CONFLICT PREDICTION AND RESOLUTION

In the conflict-prediction function, the computer continuously examines each track stored in its memory, as well as the projected future location of such tracks at a predetermined distance ahead to determine possible conflicts. These tracks may be related directly to designated airways or air routes, and in addition, rho-theta or latitude-longitude coordinates delineating area navigation routes also may serve as points at which aircraft tracks can be examined for possible conflict. These conflicts could include those where one track crosses another, or one overtakes another track, or when two tracks converge. When such a conflict appears, the computer will check each track to determine if their relative positions are within a predetermined time or altitude parameter. If such is the case, the conflict will be indicated to the controller on his display and he may then take whatever action he considers appropriate. The conflict-resolution

66 *The computer entry device is an electronic keyboard that links the controller directly to the computer.*

67 *Advanced radar terminal systems (ARTS III) provide horizontal and vertical displays showing aircraft identity, altitude and speed, next to the plane's target blip.*

function may be provided after the conflict-prediction function since the potential conflict must first be identified. Once this is accomplished, the computer will search through its stored flight data to determine possible alternative routing or altitude changes which would eliminate the potential conflict. These alternatives will then be displayed to the controller for his decision and action. He may, of course, reject all the computer-suggested possibilities, and take other action based upon his experience and knowledge of the existing situation.

DISPLAY GENERATION

Another important element in automating the ATC System is the method of the data display to be utilized, involving the amount of data desired and how long it should be displayed. The objective of the display system is to present the right amount of data, at the right place, in the correct format, at the right time, and for the right length of time, together with providing the simplest and most rapid means of communication between controller and computer. As indicated previously, a cathode ray tube (also called a plan-position indicator — PPI or "scope") is used as the display medium and provides a usable display area of about 20 inches in diameter.

Data pertaining to a selected air traffic environment are presented to the controller in the form of "data blocks." These blocks are related to a specific aircraft's radar target on the display by means of a "leader" — a line connecting the target with the data block. "Sterile areas" are provided on the displays on which no radar data are shown. These sterile areas are used to display miscellaneous flight data such as proposed arrivals/departures and weather data.

A "track ball" which is free to rotate in any direction controls the position of an electronically generated position marker on the display. The position marker moves in a direction corresponding to the direction of movement of the track ball and is used by the controller to "tag" the position of a selected target for transfer into the CCC (central computer complex). In this manner the controller can "call up" any data stored in the CCC with respect to a specifically radar-identified aircraft.

The display system accepts the processed data from the CCC and generates alphanumeric, symbolic, and map data properly positioned on the PPI. Keyboards, data entry devices, and control panels used in the console permit the controller to enter messages into the CCC and other interface equipments, and to select and control the types of data to be displayed for his use. The display system, thus, is the man/machine interface in the automated ATC System. With automation playing an increasingly significant role in the ATC System operation, the controller is able to devote his attention more productively to his fundamental role of air traffic supervision and management.

AUTOMATIC COMMUNICATIONS

Communications workload, both by pilot and controller, constitutes a major problem area in the ATC System. This workload can be so high that, especially during IFR operation in terminal areas, the pilot may be distracted severely from attention to his flying duties. On the ground, the communications workload involved in navigating and separating aircraft by radar vectoring may take 50 to 80 percent of a controller's time. Consequently, a radar controller can handle a limited number of aircraft under active control at any time, depending upon the degree of automation employed. This number might vary between a low of four aircraft under less than optimum conditions, to a maximum of perhaps 12 to 15 aircraft under optimum conditions. While the communications problem with conventional (CTOL) aircraft is significant, it becomes even more severe with vertical and short takeoff and landing (V/STOL) aircraft due to their higher landing and takeoff frequency. Automation in communications thus is an essential part of total automation programming.

Basic Features

The heart of an automatic communications system provides for the exchange of information in the form of coded, nonvoice digital signals, between ground ATC facilities and the aircraft. While the United States Universal Air-Ground Digital Communication System contemplates a transmission speed in the VHF/UHF band of

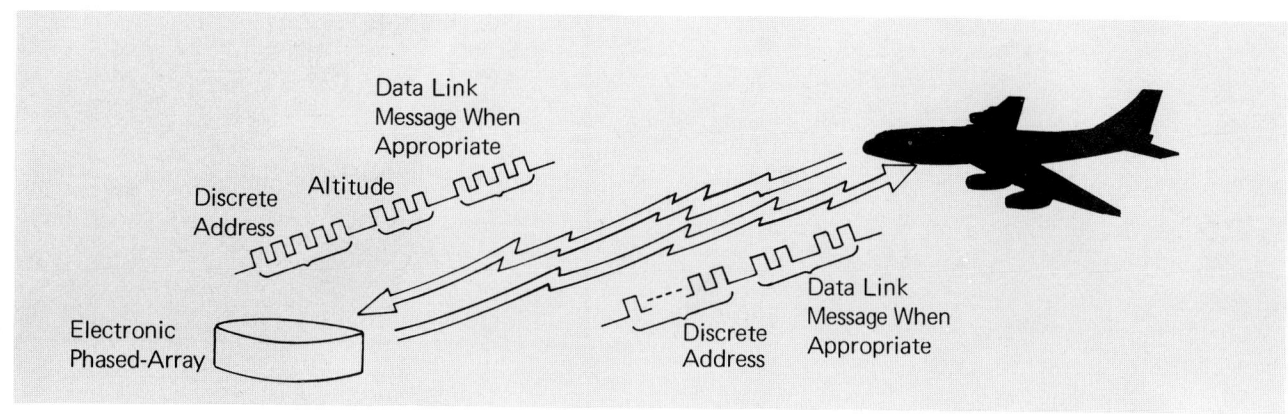

68 *High-speed digital "data link" facilities provide automatic communications between pilot and controller.*

1,200 bits per second initially in order to avoid conflict with existing ground facilities, future systems of the type being considered in automatic ATC communications development may have a bit rate up to perhaps 24,000 bits per second. (ATC voice communications, on the other hand, have a bit rate of only around five bits per second.)

An average-voice ATC message exchanged at the hurried rate of 200 words per minute requires approximately ten seconds for its transmission and acknowledgment. The digital equivalent at the planned rate of 1,200 bits per second would require perhaps one-eighth of a second. Higher bit rates would, of course, increase message-handling capacity proportionately. In addition to their greater message capacity, digital transmissions are much more reliable than voice communications as they can function with a lower ratio of signal-strength to noise.

Since digital transmissions are so rapid, large numbers of aircraft could be served by one radio channel, permitting much better utilization of existing frequencies as air traffic increases in volume.

Basic features of an automatic digital ("data link") communication system provide the following.
- Discrete address for ground-to-air and air-to-ground contacts.
- Acknowledgment of a discrete address contact automatically by the receiver equipment.
- Emergency signaling for transfer to pilot/controller voice-communications channel.
- Self-checking capability for system error.
- Instant alarm feature in case of system failure.

Voice communications will continue to be used in the ATC System for the following.
- Aircraft not equipped with automatic digital communication devices.
- Pilot/controller consultations, nonroutine or unusual messages.
- Backup in the event of any partial digital communication system failure.

Air/Ground Communications

The first application of automatic air/ground communications is in position reporting. This, in effect, is the function of the ATCRBS or secondary surveillance radar system (SSR). However, other forms of automatic position reporting may be achieved, such as by transmitting in digital form the position of the aircraft as derived from the on-board navigation system.

In a more direct air/ground automatic communication mode, the pilot will be able to transmit the following types of communications to the ground ATC System.

- Request for ATC instructions: Taxi, takeoff, landing, track, altitude, and speed.
- Requests for changes to existing ATC instructions: Routings or tracks, headings, altitude, and speeds.
- Requests for communications: Change frequency and change to voice frequency.

Ideally, in an advanced ATC automation environment, in addition to sending the above types of communications, the pilot should be able to file a complete flight plan through his digital air/ground communications equipment. This flight plan could then be the basis for receiving further automatic ground/air ATC instructions.

Filing a complete flight plan via air/ground automatic communication equipment would merely be an extension of the procedure to send requests for ATC instructions or amendments. Normally, this capability would not be used by airline or other scheduled operators. It would be very useful, however, to the large numbers of general aviation aircraft (CTOL and V/STOL) which will be operating in the future from takeoff points not served by conventional flight-plan filing facilities; also for in-flight operations when the pilot desires to enter protected airspace.

Ground/Air Communications

In ground/air automatic communications, the controller — by his own decision or by computer-generated assistance — transmits to the pilot, via an appropriate carrier, high-speed digital messages conveying instructions in the following categories.

- General: Authorization to proceed as per filed flight plan; specific departure, enroute, and arrival tracks or routings to be followed; taxiway and runway to be used; air-to-air separation instructions; proceed-hold.

CHAPTER VIII

NAVIGATION

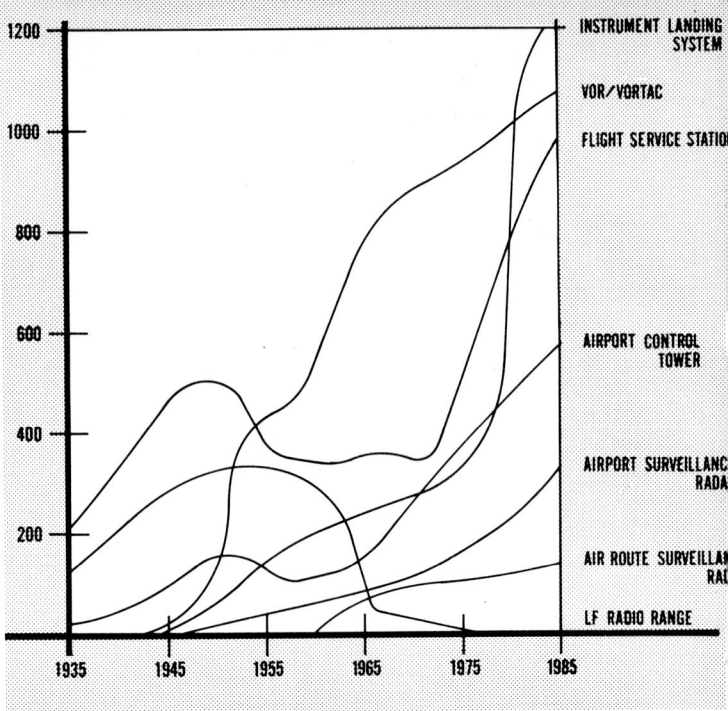

69 *Quantitative trends in navigation and associated facilities used in the control of air traffic (USA).*

- Headings: Specific headings to be followed; heading changes.
- Altitudes: Altitudes to be observed at specific points; cruising; altitude changes.
- Speeds: Speed to be observed during climb, enroute, approach; speed changes.
- Communications: Automatic digital communication frequency to be used; frequency changes; request for voice communication and frequency.

The foregoing types of communications may be classified as "priority ATC" and would not be mixed on the same channels with any other type of messages. They will lend themselves to the use of a relatively simple electronic readout device in the cockpit showing alphanumerics and symbols. The readouts will convey the appropriate instructions to follow until new or changed instructions are received.

Air navigation is to the control of air traffic as highways are to the control of automotive traffic and rails to the control of railroad traffic. Highways have developed into multilane traffic channels with cloverleafs, underpasses and overpasses to separate and facilitate the movement of private automobiles, bus and truck traffic moving in different directions and at different speeds. Aircraft pilots must depend on *navigation* to provide the same concept for the control of air traffic — multiple tracks, overpasses, underpasses, and cloverleafs in the sky — at various altitudes from the ground on up to 20 miles high (or more). A track in the sky to be followed by an aircraft must be just as definite as a highway on the ground. The pilot must be able by navigation to follow these tracks precisely at all times under all weather conditions and at all speeds — from zero forward speed in the case of vertical takeoff and landing aircraft (VTOL's) up to several thousand miles per hour in the case of supersonic aircraft (SST's).

Following these tracks, however, is not enough; the pilot also must be able to determine accurately his position at all times along the track. The accuracy with which pilots can carry out these tasks obviously affects the ability of the Air Traffic Control System to provide for the safe separation and efficient spacing between aircraft as well as to expedite the flow of air traffic. Thus, an important factor in modern air-navigation technology is its capability to provide a high degree of accuracy to

the pilot in the utilization of airspace for *air traffic control purposes.*

Basically, navigation of aircraft is the responsibility of the flight crew — not that of the air traffic controller. The controller's job primarily is to provide for the expeditious flow of air traffic and to prevent collisions between aircraft — not to navigate aircraft, except perhaps in case of an emergency.

Air navigation is such an important factor affecting the capacity and efficiency of the ATC System that it must be treated in depth in order to consider solutions to some of the fundamental problems involved in the control of air traffic, particularly in high-density, mass air-transportation environments. Accurate navigation capability is needed not only so that pilots will know constantly their positions and be able to follow precise paths; it also is a necessity in providing the basis for the controllers to know that pilots can follow accurately designated routings and tracks without the need for constant radar surveillance and vectoring by the ground-control system.

There are several methods of navigation which are used in the ATC System. Some of these rely on ground-based radio navigation aids. Others use self-contained airborne systems which can operate independently of ground aids.

BEARING/DISTANCE NAVIGATION

In this concept, more or less fixed "airways" are defined

70 Basic VOR/DME (VORTAC) navigation facility.

by means of radio stations strategically located on the ground to delineate a prescribed routing to be followed in the ATC System. These airways have specified widths — in the United States generally 8 miles — and may be classified as "high altitude" for flight above a specific height above sea level (such as 12,500 or 18,000 feet), or "low altitude" for flight below the demarcation level. Different degrees of air traffic control are applied to these two distinct airway structures.

VOR/DME Navigation

This system of air navigation uses equipment in the aircraft and on the ground which provides the pilot with indications as to distance (rho) and bearing (theta) from specific VOR/DME ground stations. The VOR/DME type of rho-theta navigation is recognized by the International Civil Aviation Organization as the international standard "short-range navigation aid." Extensive VOR/DME coverage is provided in the continental United States, Canada, Western Europe, and in parts of Latin America, the Far East, and Middle East. It is expected that this navigation system will continue in widespread use, at least until the end of this century, although other more modern navigation systems are being introduced.

The United States developed the VOR (very-high-frequency omnidirectional range) originally in the 1940s as a modernized replacement for the four-course, low-frequency ranges which marked the advent of ground air navigation aids in the late 1920s and early 1930s and which served as the basic air-navigation system during the first ATC generation. The VOR navigation system provides the pilot with radial bearings (azimuth) throughout 360° going to or from the ground VOR station. An infinite number of such "radials" can be used for navigation. During the 1950s, the United States Department of Defense developed an ultra-high-frequency navigation system for military aircraft called TACAN (TACtical Air Navigation), which gives both azimuth and distance to suitably equipped aircraft. When this facility is installed at a VOR location, the combined facility is called a VORTAC, with the distance measuring portion being common to each. The azimuth portion is separate. (A low-powered VOR installation designed to serve a terminal area in the vicinity of major airports is designated a TVOR.) The advent of the VORTAC navigation system in the 1950s was another second generation ATC System landmark along with radar.

In the rho-theta navigation system, a distance measuring equipment or DME, which is common to the TACAN equipment, is provided giving the pilot a continuous reading of the slant range distance to or from the DME ground station in nautical miles. The standard DME ground station is capable of transmitting distance measurement signals to about 100 aircraft at one time, but this capacity can be expanded in high-density traffic locations. The airborne DME interrogator sends a randomly spaced series of pulse-pairs which are replied to by the ground transponder. Measuring the time lapse gives distance to the ground DME station which is shown on a continuously registering, digital display indicator in the cockpit instrument panel.

The DME ground station — with a few exceptions in some parts of the world — is located at the same site as the VOR station. When a pilot tunes to a VOR (or VORTAC) station, his DME receiver is automatically tuned to the DME ground station associated with that VOR station. By his cockpit instrumentation, the pilot thus knows the bearing (or radial) he is on with respect to the VOR station as well as the distance of his aircraft from that station.

The VOR permits the establishment of predetermined tracks or routes for pilots to follow, within certain accuracy tolerances, as defined by VOR radials. At 100 miles, the effective width of a radial can be in the order of 8 miles; accuracy in resolution, however, increases as the VOR station is approached. This characteristic of the VOR has the effect of squeezing air traffic into single-lane "precision collision courses" as aircraft fly toward a ground station. Approximately 20 percent of near midair collisions reported during VFR conditions occur in the vicinity of VOR sites.

As a result of such channeling of air traffic when VOR/DME (or VORTAC) is used as the basic navigation aid, ATC has had to rely on altitude separation between aircraft flying in opposite directions or in overtaking situations,

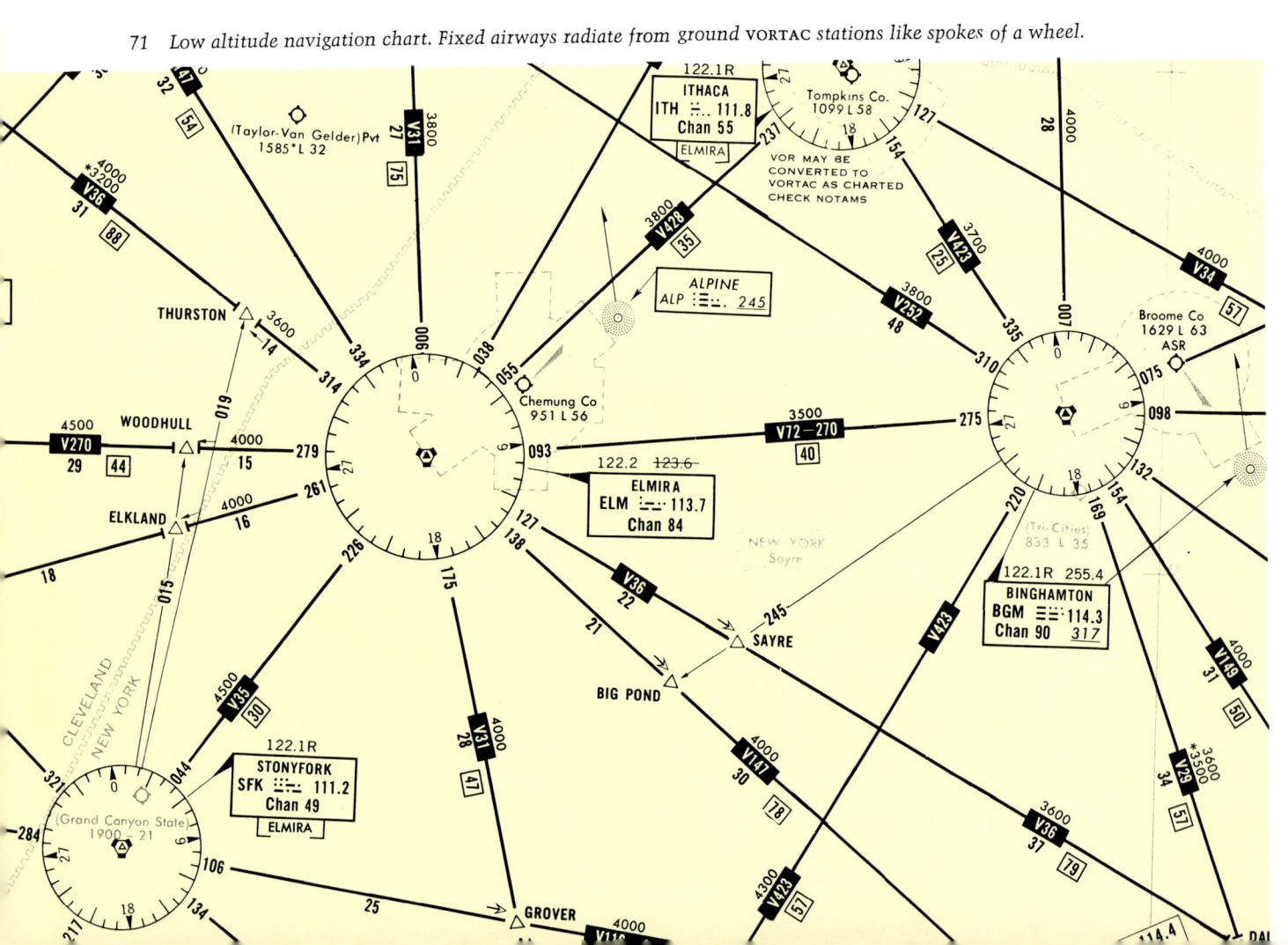

71 Low altitude navigation chart. Fixed airways radiate from ground VORTAC stations like spokes of a wheel.

rather than lateral separation. This has proven wasteful of airspace due to the inability of the VORTAC navigation system to permit pilots to follow parallel or multiple tracks on their own responsibility within the same general airspace environment. Such deficiency has to some extent been compensated by the controller's use of ground radar to vector aircraft as described earlier. In effect, the controller thus navigates the aircraft from the ground (with high controller/pilot communications workload) in order to achieve lateral, or longitudinal, separation from other aircraft in conflicting traffic situations.

Standard VOR ground-system accuracy is typically ±1.9° to ±2.0° of the indicated bearing. Similarly, DME accuracy is typically ±.5 nm or 3 percent of the measured distance, whichever is the greater. TACAN systems afford approximately equivalent accuracies. Improved overall-system accuracy can be achieved by installation of Doppler VOR (DVOR) and precision VOR (PVOR) ground stations. PVOR, for example, can provide a bearing accuracy of approximately ±.3°. Precision DME (PDME) ground stations can have a potential accuracy in the order of 200 feet.

The VOR/DME (TACAN) navigation system is subject to line-of-sight limitations which in effect means that its navigation signals are not received at low altitudes down to the surface unless the receiving aircraft is in close proximity to a ground station.

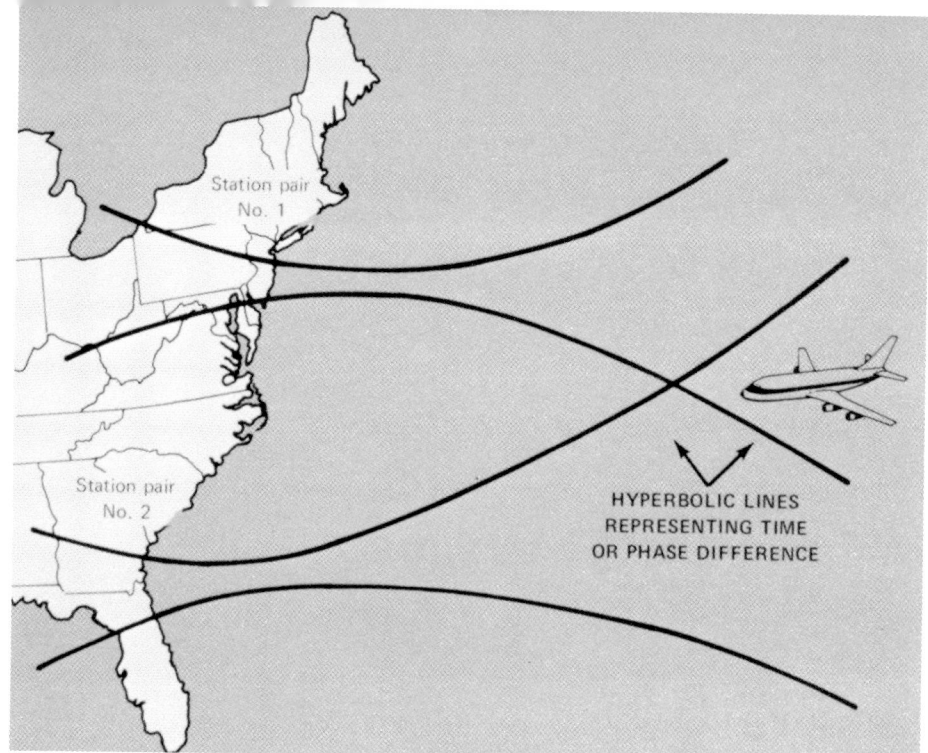
72 Position fixing using a hyperbolic navigation system.

Direction Finding

Direction finding or "DF" is one of the oldest methods of navigation. However, it basically provides only a measurement of bearing relative to the aircraft or ground station (unless sophisticated auxiliary computers are used). DF originally was developed in the 1920s to provide a "fix" on an aircraft's position as determined by the intersection of lines of bearing taken by two or more ground direction finding stations on the aircraft's radio signals. It was a primary navigation system used in some parts of the world until the mid-1940s.

As an auxiliary navigation tool, DF is employed principally at airports. The DF readout equipment (usually installed in a control tower), shows a line of position from the airport on a special CRT (cathode ray tube) display or it may be included with the normal ATC radar display. The bearing is taken on the aircraft's voice communication transmitter. An actual position, or fix, is possible only if two or more bearings are plotted simultaneously from different ground stations.

Generally speaking, this is an emergency type of navigational device which enables a controller to help a "lost" pilot find an airport by giving him headings to follow. It may also be used to verify radar plots by integrating the DF lines of position (identified by voice communication) on an ATC PPI. The technique is useful on occasion in a search and rescue operation to plot the location of an aircraft down at an unknown position.

Automatic Direction Finding

Automatic direction finding (ADF) is an airborne counterpart of the ground DF concept. In an aircraft, the ADF permits taking continuous bearings automatically on any ground station operating on a frequency usable by the airborne equipment. The ground stations which are available to the pilot for this purpose include a number of "compass locators" in terminal areas, nondirectional beacons (NDB) and commercial radio-broadcasting stations. All of these may be used by the ADF-equipped aircraft for enroute navigation and for approaches to certain airports in remote areas. ADF is used mainly as a "backup" navigation device, or in areas of the world where traffic is light and ground navigation aids are limited. It does not provide the position-fixing accuracy or display needed for modern air navigation in high-density traffic patterns.

HYPERBOLIC NAVIGATION

The hyperbolic navigation technique relies on a combination of a "master" and one or more "slave" ground radio stations which generate radio signals forming sets of hyperbolic lines from each master-slave pair measured as a time (or phase) difference in the reception of signals from the master and slave stations. The intersection of any two hyperbolic lines provides a position fix which is interpreted for the pilot according to the particular hyperbolic navigation system used. Several different types of hyperbolic navigation systems have been developed which are significant from the standpoint of air traffic control. Hyperbolic navigation systems permit reception down to the surface due to the propagation characteristics of the low (LF) or very low (VLF) frequencies which they employ. Accuracies are affected by diurnal and seasonal variations, as well as by atmospheric interference.

Decca

The Decca navigation system consists of a master and three outlying slave stations to form a "chain." The three slave stations are positioned about 70 miles from the master station in a triangular pattern with the master station at the center, although neither the distance nor the geometry is critical. The stations each transmit on a different frequency which is some multiple of a certain fundamental frequency in the order of 15 kHz. Decca is a continuous wave (CW), phase comparison, hyperbolic system. It utilizes an airborne computer to determine position by relating phase measurement to transmission time. It has a greater coverage at night than it has during the daylight hours, the nighttime coverage averaging 200-250 miles.

From the computer output, the pilot continuously monitors his track by means of a pictorial display. This display is designed to work with any navigation system, including VOR/DME, Doppler, Inertial, LORAN C, as well as with the Decca hyperbolic chains. The accuracy of the presentation depends, of course, on the accuracy of the navigation data fed into the computer.

Dectra

This is a hyperbolic navigation system for use over large bodies of water, such as the North Atlantic. A standard computer and pictorial display equipment also may be used with the Dectra system. The system was initiated in the North Atlantic area in 1957. In general, the Dectra system provides a number of parallel tracks, running east/west between Newfoundland and the United Kingdom, although north/south tracks are available.

Harco

In the early 1960s, Eurocontrol expressed an opinion to the effect that, although VOR/DME would be used in Western Europe during many years to come, it might be necessary to implement a new or an improved area navigation system. Preliminary specifications for such a system were worked out by Eurocontrol and were distributed to all major European electronic firms. The HARCO (Hyperbolic Area Coverage) system was the outcome of the joint development of three of these firms. A determination of action regarding the selection and implementation of the appropriate system is the responsibility of ICAO.

LORAN

This hyperbolic navigation system was developed by the United Kingdom in the early 1940s, initially for aviation; however, it served extensively for United States naval navigation during the latter part of World War II, but was used only to a slight extent by aircraft.

The original LORAN (LOng RAnge Navigation), called LORAN A, was retained in operation after World War II primarily by the United States government which provided partial coverage over the North Atlantic, the Gulf of Mexico, and the Pacific Ocean.

LORAN systems are pulse rather than CW-type systems such as DECCA. Aircraft position fixing with LORAN A is effected by measuring the time delay between reception of synchronized pulse transmissions from the ground stations. With LORAN C and D, a synchronized carrier frequency is also transmitted.

The LORAN A system is generally considered obsolete in comparison with modern standards of air navigation. An improved version of this system is LORAN C system, developed around 1960, which consists of a master and two or more slave stations operating in the LF band with usable ground-wave signals up to 1,500 nautical miles landward and 1,900 nautical miles seaward. Readout of navigation data is by continuous numerical display of two time-difference readings (hyperbolic lines of position). These readings are then interpolated on a LORAN C navigation chart to determine position fixes. LORAN C transmitting chains cover the Atlantic Coast of the United States, the North Atlantic, the northern European area, the Mediterranean Sea, and the northern and central Pacific Ocean regions. "Differential" LORAN C can provide increased accuracy through computerized compensation in a specific area for local diurnal, seasonal, and other error-contributing factors.

A further hyperbolic navigation development — LORAN D — provides short-range coverage. The LORAN D ground installation consists of one master and two slave stations for each chain. The LORAN D receiver can be used with a computer to provide numerical readouts of aircraft position (latitude and longitude), bearing and distance to destination, time to destination, and track error and correction.

Omega

The Omega system is a member of the hyperbolic navigation family operating in the VLF band. A long-range navigation system, Omega is intended for use by ships, aircraft, and submarines. Complete global coverage (over land and sea) requires eight stations at strategic points on the globe. Signal coverage is available down to the surface in most areas. "Differential" techniques also may be applied to Omega to improve localized accuracy.

Operation of the Omega system is under the direction of the United States Department of Defense, primarily for naval use, but the system is available for use by aircraft of any nationality. Airborne receiving equipment is available commercially in various models.

Basically, the aircraft's location is established in the Omega system by the intersection of two hyperbolic lines of position. These lines of position are determined by measuring the phase difference of two of the received signals (slaves) in relation to a third signal (master). Read-

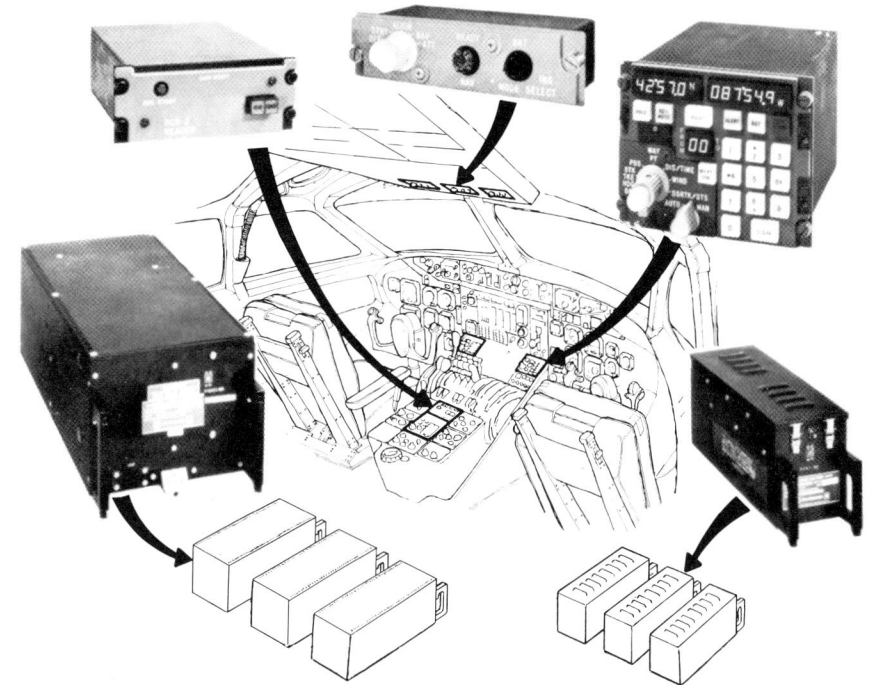

73 *Inertial system components and equipment distribution.*

out in the naval application is in hyperbolic coordinates (requiring plotting of the fix on a chart). In airborne use, a computer is involved which permits automatic numerical readouts of latitude and longitude, course to steer, distance to go to a selected point, distance off course, and other similar navigational data.

Hybrid

Hybrid applications are possible with combined inputs from different types of VLF facilities, such as Omega ground stations and high-powered United States naval communications stations. Using a suitable airborne computer with appropriate software, a pair of VLF stations located to the left and right of the intended track provide steering information, and another pair ahead and behind provide along-track position information, giving "global" navigation capability.

INERTIAL NAVIGATION

Inertial navigation systems (INS) can be used in the air, on the surface, or underwater (inertial navigation guided the USS *Nautilus* in its undersea arctic voyage of 1958). It is used in most jet aircraft — airline, business, and military. In simplest terms, inertial navigation is "integrating acceleration to determine velocity and position."

What this really means is that an aircraft, for example, equipped with inertial guidance needs no contact with the outside world after takeoff. Navigation can be carried out entirely by measurements within the vehicle, without reference to any exterior source of information, on or off the earth. No radio signals are needed from ground stations, no radar navigational instructions are required. The inertial system "knows" where it is at all times and its flight path can be controlled by the pilot to any desired destination, simply because the aircraft's inertial navigation equipment continuously computes where the aircraft has gone from the known point from which it took off.

The three primary advantages made possible by an INS are: an accurate aircraft attitude reference which is unaffected by acceleration or aircraft maneuvers; a heading reference of extremely high accuracy; and a self-contained global navigation system.

The development of instrumentation for inertial guidance involves mathematics of a very advanced order. The real "breakthrough" was made possible following the introduction in the 1960s of extremely sophisticated gyros and accelerometers, as well as computers of a size feasible for airborne installation. Nevertheless, there is still one instrumental problem in inertial guidance technique which has not been completely solved. This is, that, in the current state of the art, gyros are not perfect. Drift, due to inertia and internal friction, accumulates at the rate of one to two nautical miles per flight hour (development of gyros involving laser beams, gaseous bearings, and other design improvements may reduce this problem). In the event that this drift rate results in an unacceptable degradation of navigational accuracy in a given operational environment, the pilot must "update" the INS (i.e., correct drift error) by reference to other position fixing means.

The inertial navigation display gives the pilot numerical readouts on the following data.

- Present position in latitude and longitude.
- Ground speed and present track.
- Wind direction and speed.
- Distance from desired track.
- Correction required to correct track error.
- Distance/time/bearing to next checkpoint and/or destination.
- Pitch and roll attitude.

Before takeoff, the pilot aligns the system by inserting into the computer his known latitude and longitude, plus the latitude and longitude of his en route checkpoints. To update the inertial navigation system, the pilot presses an "update" button on the control panel and "freezes" the display, while the computer goes on operating. Latitude and longitude of a new fix — obtained from ground radio aids — may be compared with that frozen in on the control panel. Should there be a difference, the old storage data can be changed to agree with the new fix data; and after the update button has been released, the computer will begin to use the new values.

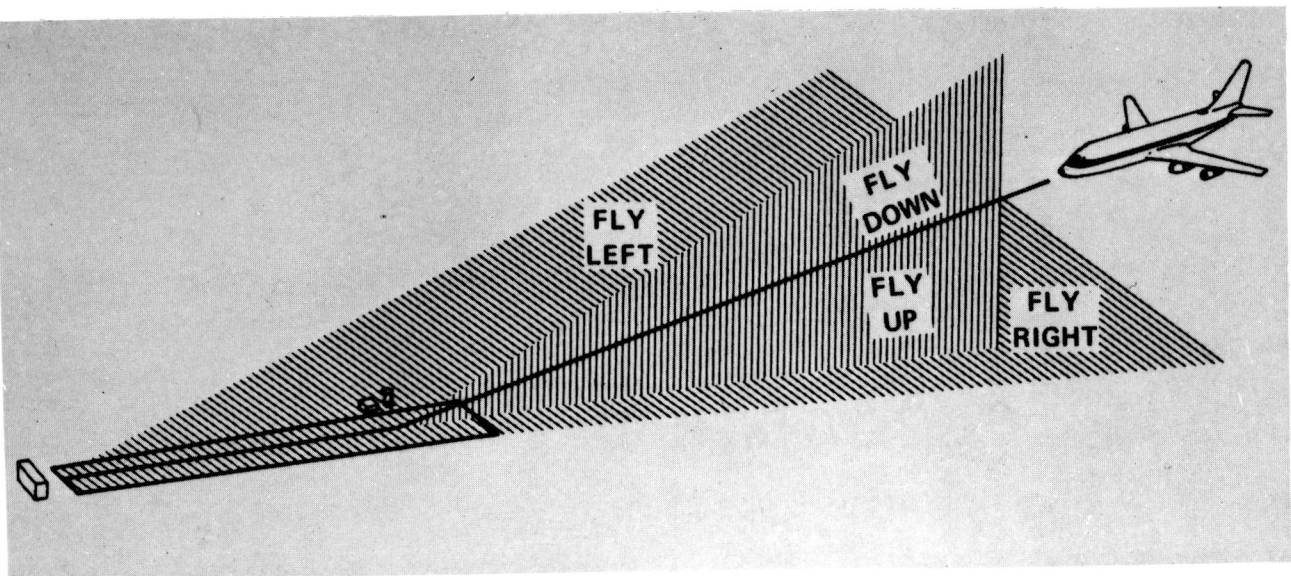

74 Conventional instrument-landing-system guidance pattern.

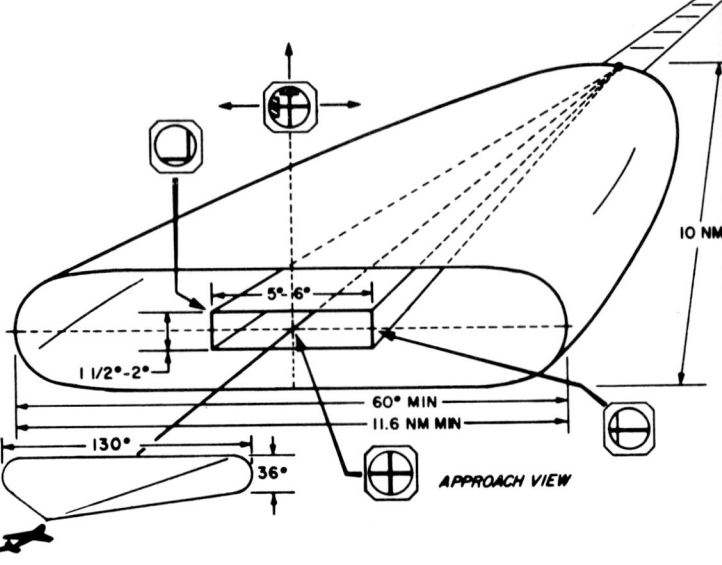

75 Fixed-beam microwave landing system.

DOPPLER NAVIGATION

This self-contained system of navigation is based on a principle discovered by Austrian physicist Christian Johann Doppler in the early 1800s. A common manifestation of the Doppler effect is in the case of the change in frequency or pitch of a train whistle as heard by an observer standing beside the track as the train speeds by. The frequency (pitch) of the disturbance—in this case, sound waves emitted by the whistle — is observed to be higher as the train approaches and to be lower as the train recedes, than the frequency of the same whistle as heard when the train is stationary.

The Doppler effect takes place at radio frequencies, as well as at audio and light frequencies. In aircraft-navigation adaptation of the Doppler effect, a frequency-modulated radio transmitter is used, operating in the VHF spectrum. The surface ahead of and behind the airplane is "painted" by beams of energy radiated from the transmitting array. Fixed or rotatable antennas may be used. The receiving arrays momentarily pick up the reflected energy from an area forward and to the right, and from an area aft and to the left. They are then switched to pick up, momentarily, the reflected energy from an equal area forward and to the left, and aft and to the right. The switching of receiving arrays is continuous while the system is in operation and takes place at the rate of one complete cycle in approximately one and one-half seconds.

A typical example of a Doppler air-navigation equipment designed for civil aircraft provides a velocity and steering indicator (VSI) which at all times displays the following.
- Aircraft track made good.
- Required track and distance to the selected destination.
- Track-error angle.
- Ground speed in knots.
- Distance scale range.
- Doppler condition warning.

Additional outputs of track-error angle, bearing and distance to destination, and present position coordinates are available for the automatic pilot and other equipment. Other versions of Doppler equipment provide digital readouts and pictorial displays.

Prior to takeoff, departure point and destination (or checkpoint) coordinates, local magnetic variation, and the surface wind vector are inserted into the airborne computer via the control indicator. The control indicator and the velocity and steering indicator are the only cockpit instruments required. Once the aircraft has left the ground, the VSI displays the selected destination relative to present position, and the control indicator shows present position in east-west and north-south distance-to-go coordinates relative to starting point. The pilot may at any time select or reset destinations on the control indicator, or instantly determine the computer local wind vector on the VSI. If loss of the Doppler signal occurs, the system automatically reverts to a memory (dead reckoning) mode of operation. The Doppler system may be updated by the pilot at any time; quite frequently it is mated with other navigation systems.

INSTRUMENT LANDING SYSTEMS

Instrument landing systems provide navigational guidance for landing derived from radio signals transmitted from ground-based electronic aids located on an airport. For accurate use, these facilities are associated with a specific "instrument" runway and are located so as to serve the touchdown point on that runway. Ground-based electronic landing aids are classified in two broad categories: ILS and MLS.

ILS

The basic ICAO international standard navigation aid for landing at airports is known simply as ILS — instrument landing system. The ILS provides guidance to the pilot in both the horizontal and vertical planes with respect to the "instrument runway" at an airport. Along the ILS approach path, VHF fan markers (sometimes co-located with LF nondirectional beacons) are provided at two or three positions to supply the pilot with spot-checks as to his distance from the runway end. These are called outer marker (4-7 miles from runway end), middle marker (about 3,500 feet from runway end), and inner marker (±1,000 feet from runway end). An improved installation incorporates a low-powered DME (distance measuring

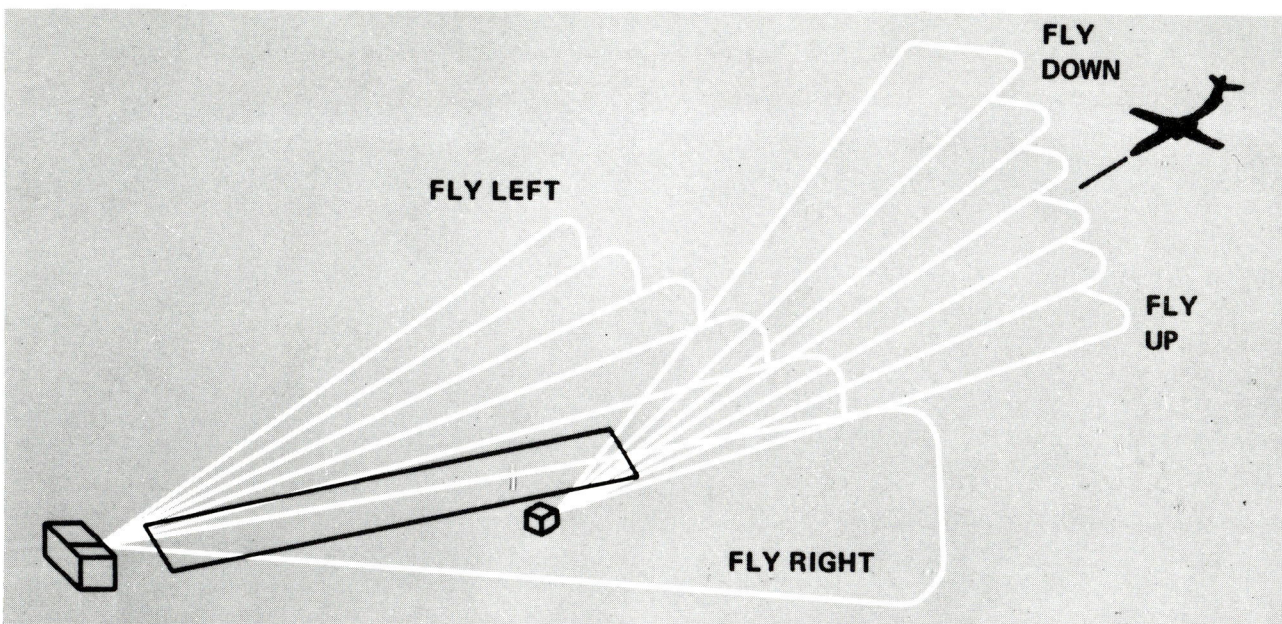

76 *Scanning-beam microwave landing system.*

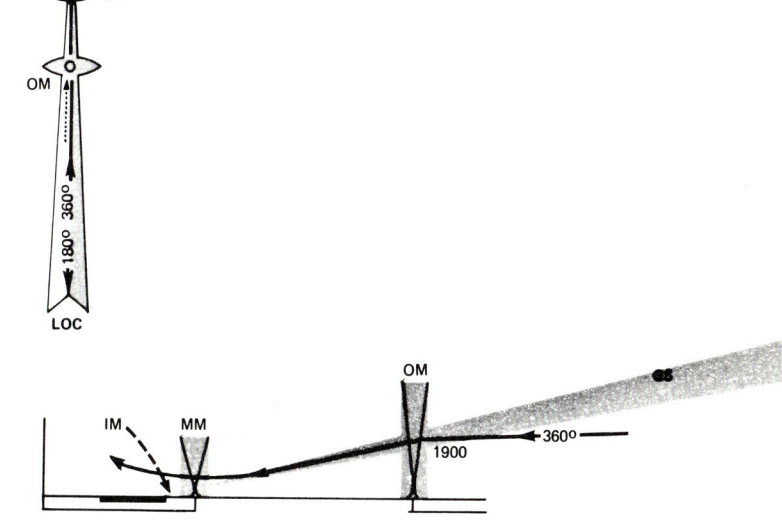

77 *Horizontal and vertical profiles using a conventional instrument landing system.*

78 *"Nonprecision" instrument approach procedure chart.*

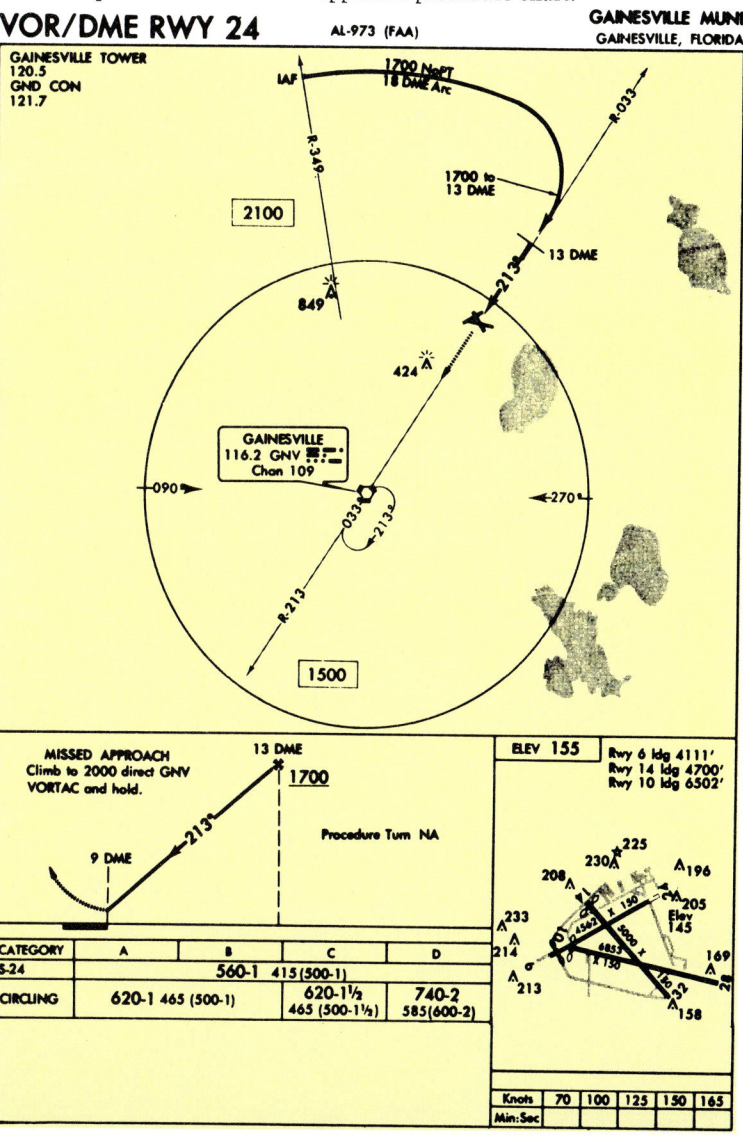

equipment) at the end of the runway to give the pilot *continuous* measurement of position along-track throughout the entire approach.

Azimuth guidance in the ILS is provided by means of a "localizer" which transmits a narrow-width horizontal radio beam aligned with the runway. Vertical guidance is provided by a "glide slope" which establishes a fixed angle of approach between 2.5° and 3° above the runway plane.

MLS

A modification of the standard ILS is the MLS — microwave landing system. Operating in a much higher frequency spectrum than the VHF/UHF ILS, advantages include reduced siting problems and, in turn, lower acquisition/installation costs, as well as mobile capabilities.

In the basic form, the MLS provides a single azimuth and glide-slope output (adjustable on the ground as to angular characteristics). In the more advanced mode — the scanning-beam MLS — variable approach courses and glide-slope angles to the end of the runway are selectable at the pilot's election.

Pilot guidance is the same with all instrument landing systems. The pilot is given command guidance (autopilot coupling optional) by reference to a cross-pointer instrument which combines left-right (localizer) and up-down (glide-slope) indications. The accuracy of these systems is classified in different categories (categories I, II and III) which in turn determine the landing weather minimums under which they can be used at a particular airport.

BASIC INSTRUMENT-APPROACH PROCEDURES

Instrument-approach procedures (IAP's) are of two basic types.
- Precision-approach procedure in which a ground-based electronic aid is provided.
- Nonprecision-approach procedure in which no ground-based electronic glide slope is provided.

For precision approaches, an ILS, MLS, or PAR is used. Nonprecision approaches use a VOR, VOR/DME (VORTAC) or ADF/NDB as the guidance source.

In an instrument approach the pilot transitions from en route flight to an "initial approach fix" (IAF) at a predetermined safe altitude. He then descends to a "final approach fix" (FAF) at an established distance from runway end, at which point he commences his final approach for a landing. Each established IAP serves only a specified runway.

Landing minima are determined by the appropriate authority — FAA in the United States — for each IAP. In the case of nonprecision-approach procedures, these are identified by an MDA (minimum descent altitude) which is the lowest mean-sea-level (MSL) altitude to which descent is authorized on final approach. For precision-approach procedures, the landing minima are identified by the DH (decision height) which is the height above sea level at which the pilot must make a decision

79 *"Precision" instrument approach procedure chart.*

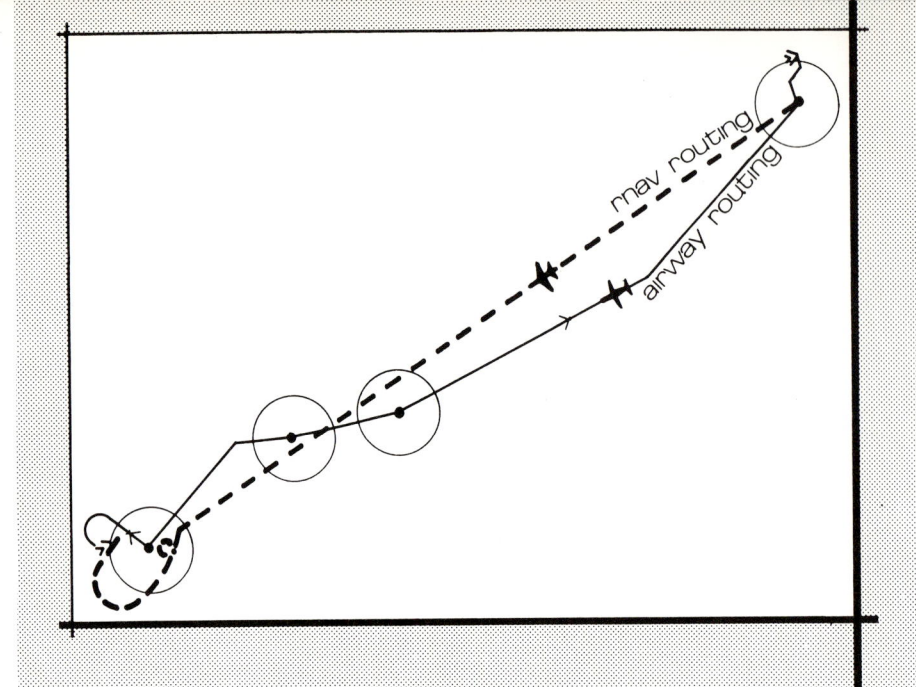

80 *Area navigation provides direct RNAV routings without the need to overfly each ground VORTAC as is the case of airway routings.*

to either continue the approach or execute a missed approach.

Whereas all precision approaches are "straight-in" to the runway, nonprecision approaches may be straight-in or "circling." HAA (height above airport) indicates the height of the MDA above the airport elevation and is published in conjunction with circling minimums. HAT (height above touchdown) is the height of the DH or MDA above the highest runway elevation in the touchdown zone (first 3,000 feet of runway). HAT is published in conjunction with straight-in minimums.

Landing minima for nonprecision approaches generally are higher than for precision approaches. They also are higher for circling approaches than for straight-in approaches. Variations between different categories of aircraft also may affect landing minima.

Instrument-approach procedures play an important part in the control of air traffic as they significantly affect traffic routing and airspace occupancy. Controllers must be thoroughly familiar with all IAP's for airports within their areas of jurisdiction.

AREA NAVIGATION

As contrasted to the constraints imposed by the requirement to navigate from one ground-navigation station to another on a fixed airway system, area navigation (RNAV) permits aircraft operations on *any* desired course within the coverage of external navigation signals or within the limits of self-contained navigation system capability.

Area navigation airborne equipment basically provides the pilot with continuous information as to track and position along track — called 2-D RNAV (position determined in *lateral* and *longitudinal* dimensions). An added capability provides the pilot with *vertical* guidance, and this is then called 3-D RNAV. If 3-D RNAV is coupled with *time-control* capability, it is referred to as 4-D RNAV. With 4-D RNAV, the pilot, by reference to his airborne equipment, can arrive on a predetermined track at a desired point in space (or on the ground) at a desired altitude and at a desired time.

RNAV Benefits

The United States took the lead to incorporate area navigation concepts in its Air Traffic Control System with an internationally attended RNAV symposium in 1972. Prior to that time, an *Advisory Circular* on the design and use of 2-D RNAV equipment was issued by the Federal Aviation Administration in 1969, based on several years of preparatory studies and tests both by the government and private industry. In 1973, standards were developed covering 3-D RNAV, and a comprehensive program was initiated aimed at making area navigation techniques and concepts an integral part of the Air Traffic Control System by the mid to late 1970s.

Typical applications of 2-D RNAV make the following possible.

- Congested area bypass routes.

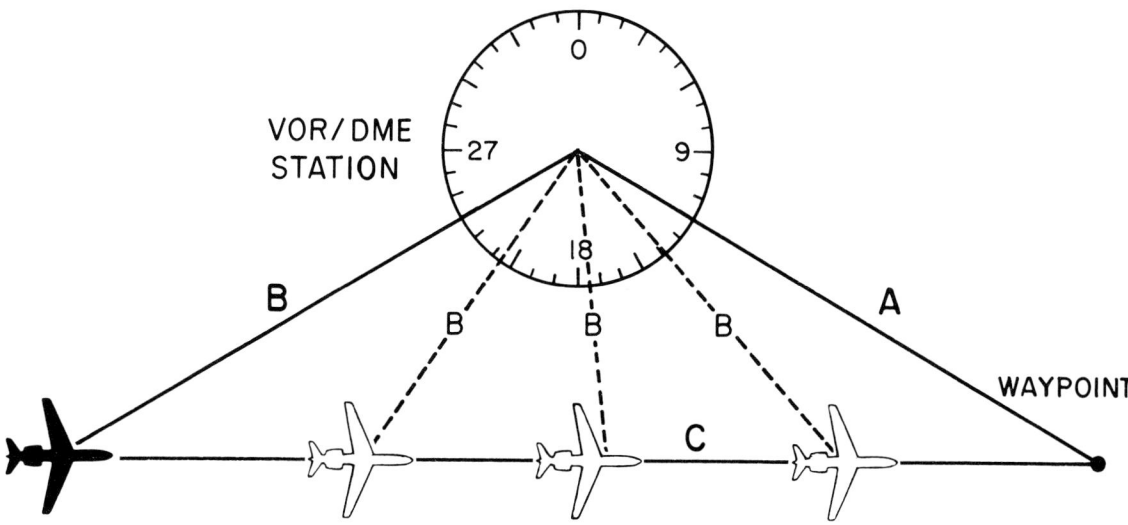

81 *Distance and bearing from ground station (A) defines waypoint coordinates; distance and bearing signals (B) are fed from aircraft's VOR/DME (or TACAN) receivers into airborne RNAV computer; track and position along track (C) are continuously computed from A and B.*

- Multiple routes and parallel tracks to allow segregation of traffic according to speed or other operating characteristics.
- Pilot navigation of commonly flown radar-vector paths.
- Improved alignment of routes to permit direct flights.
- Dual routes for one-way traffic.
- Instrument approach capability to airports/runways not equipped with an electronic landing aid.
- Optimum location of holding patterns.

With the added vertical dimension provided by 3-D RNAV, typical additional applications include these.

- Adherence in three dimensions to "standard instrument departure" and "standard terminal arrival routes" (RNAV SID's and STAR's) so as to reduce cockpit/controller communications; give pilot more time for flight management; controller more time to supervise/expedite traffic flow and monitor for conflict detection.
- Ability to make efficient altitude changes by following vertical routes (tubes) of known dimensions.
- Stabilized descent in instrument approach procedures using computed glide-slope information.
- More effective compliance with three-dimensional noise abatement procedures, including multisegment descent/ascent profiles.

In the application of 4-D RNAV, the controller assigns the pilot a specific time to arrive at a designated point in space, or at the runway touchdown point. The pilot, by reference to his 4-D RNAV system, adjusts his speed in order to perform the specified 4-D flight profile with a high degree of accuracy.

Area navigation systems fall into two general categories: those continuously requiring inputs from ground-based navigation facilities; and those which are self-contained and do not need continuous inputs from ground-based navigation stations.

Ground-based RNAV

Because of their widespread use in the United States and elsewhere, VOR/DME (VORTAC) facilities provide a readily available source of input over many land areas of the world for RNAV systems relying on ground-based navigation stations. Other RNAV systems using ground-based navigation facilities include Omega, LORAN, Decca, and hybrid systems relying on inputs from VLF/LF transmitting stations.

Some differences exist in the characteristics of these various systems. For example, the VOR/DME (VORTAC) facilities operate on VHF and UHF and, as a consequence, are not affected adversely by atmospheric disturbances. On the other hand, because of their line-of-sight signal-propagation characteristics, coverage varies with distance from the station and aircraft altitude. At distances up to 15 nautical miles from the station, as an illustration, signal coverage normally should be adequate down to 200 feet or less above the surface; at 20 miles down to about 300 feet, and so on.

In the case of VLF/LF ground stations, they provide signal coverage down to the ground. They are subject, however, to atmospheric interference and variations in coverage and accuracy due to diurnal and seasonal effects.

All systems require one or more airborne computers to process the input signals and give the pilot the desired display format. Flight-plan information can be inserted into the computer system manually, or such information can be held in a "flight-data storage unit" (FDSU) and inserted into the computers by means of an "automatic data-entry unit" (ADEU).

Although there are some variations in the operation of these different types of RNAV equipment, their use by the pilot essentially is the same. Since RNAV equipment based on rho/theta (VORTAC) inputs is most numerous, a typical system using such inputs illustrates the operation of all.

To delineate a desired flight path or locate a landing area, one or more precise locations on the surface called "waypoints" are selected by the pilot. Each waypoint is defined by its distance and bearing (azimuth) from a VORTAC station and by its latitude and longitude. Provision can be made for the storage of waypoint coordinates and other RNAV data prior to departure, either manually by the pilot or automatically from an FDSU and ADEU. Thus, a complete RNAV flight may be preprogrammed to ensure accuracy and reduce pilot workload. If changes in programmed RNAV flight data subsequently are required,

82 *Single waypoint, general aviation, 2-D* RNAV *system showing (1) distance and bearing coordinate selector; (2) track selector/indicator.*

83 *Dual-waypoint, 2-D* RNAV *system. Control head (1) permits storing coordinates for two waypoints; symbolic pictorial indicator (2) provides for track selection and readout.*

these can be accomplished manually or automatically while en route.

The horizontal flight path to or from the waypoint is determined by the track which the pilot wishes to follow. This track is set — manually or automatically — into the instrument which is to be used by the pilot for navigational guidance. These may be a "course deviation indicator" (CDI), a "horizontal situation indicator" (HSI) or a "symbolic pictorial indicator" (SPI). The first two are standard navigation instruments, and the third is a special optional instrument for RNAV use.

These instruments all give the same command guidance to pilot (or autopilot). By keeping the vertical cross-pointer centered the aircraft is on the desired track. If this cross-pointer is to the right of center, the track is to the right and the pilot steers the aircraft to the right to get on track (and vice versa).

An additional feature of a SPI is a miniature aircraft in the center of the indicator showing the relative heading of the aircraft with respect to the waypoint. Keeping the vertical pointer centered means that the aircraft is on the desired track; when the horizontal pointer intersects the vertical pointer at the center of the indicator, the aircraft is over the waypoint. A digital indicator also is provided to continuously display distance in nautical miles to, or from, the waypoint.

The basic objective of the application of an area navigation system in terminal areas is to permit pilots to follow prescribed arrival and departure flight paths, closely spaced in accordance with the most efficient traffic patterns. These patterns also may specify "standard" altitudes to be followed at particular points along the flight paths.

Many major airports and terminal areas require the designation of departure and arrival routes which are standardized and coded. The routes used for departures are known as standard instrument departures (SID's), and those utilized for arrivals are termed standard terminal arrival routes (STAR's). Such routes are established and published on charts to simplify the delivery of ATC instructions to pilots of arriving and departing aircraft. These routes, however, frequently require a considerable amount of precise navigation on the part of the pilot, resulting in significant cockpit workload. On the other hand, by using an area navigation system, a great deal of the additional workload placed on the pilot can be eliminated. RNAV SID's and STAR's also are published on appropriate charts.

By designating, in advance, terminal-area RNAV flight paths and giving each a specific identification, it is possible for controllers to supply precise routing instructions to pilots merely by stating the desired flight-path identification, which each pilot then follows on his own responsibility. With RNAV-equipped aircraft, the controller does no direct vectoring unless, by radar monitoring, he determines that a particular aircraft is not following the speci-

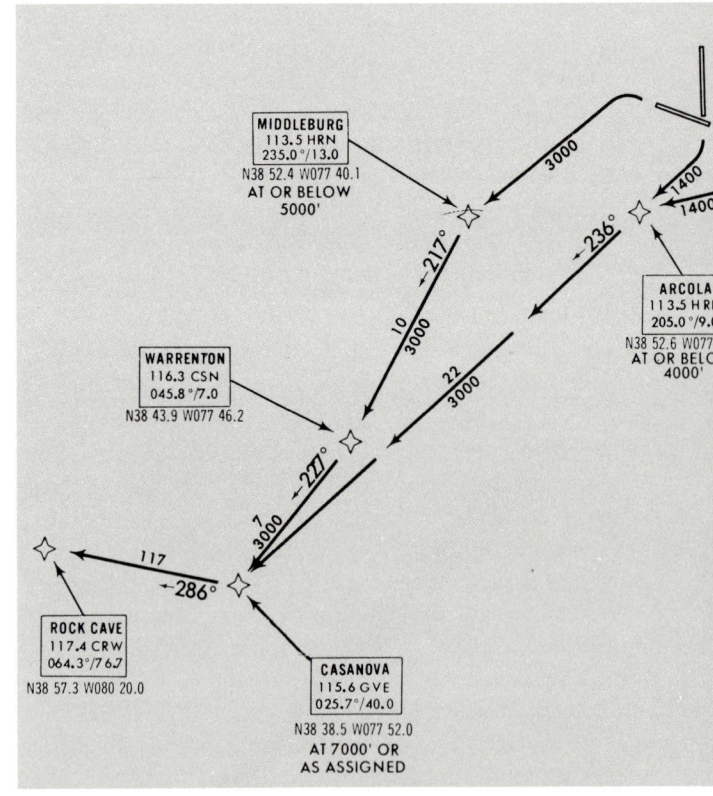

84 *An* RNAV *standard instrument departure* (RNAV SID) *procedure merely requires the controller to identify it (e.g., "Casanova one* RNAV *departure"), and the pilot then executes that procedure by reference to his airborne* RNAV *equipment.*

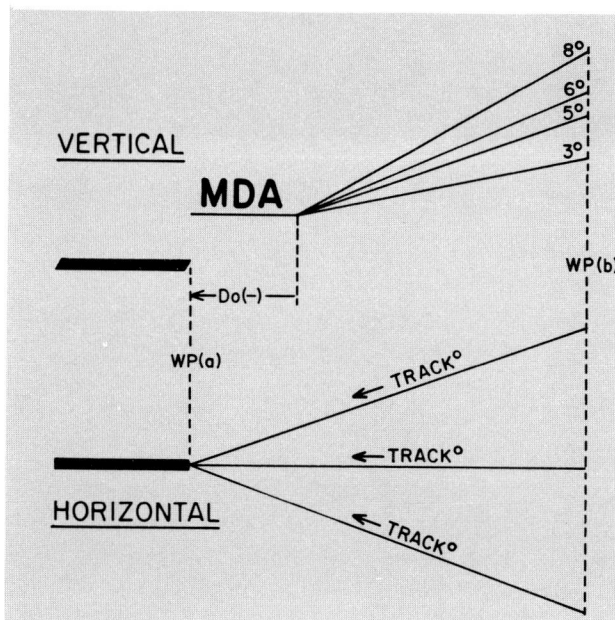

85 With airborne RNAV, *a pilot may set in the appropriate minimum descent altitude (*MDA*), and reach this altitude on a desired track and on a selected glide slope.*

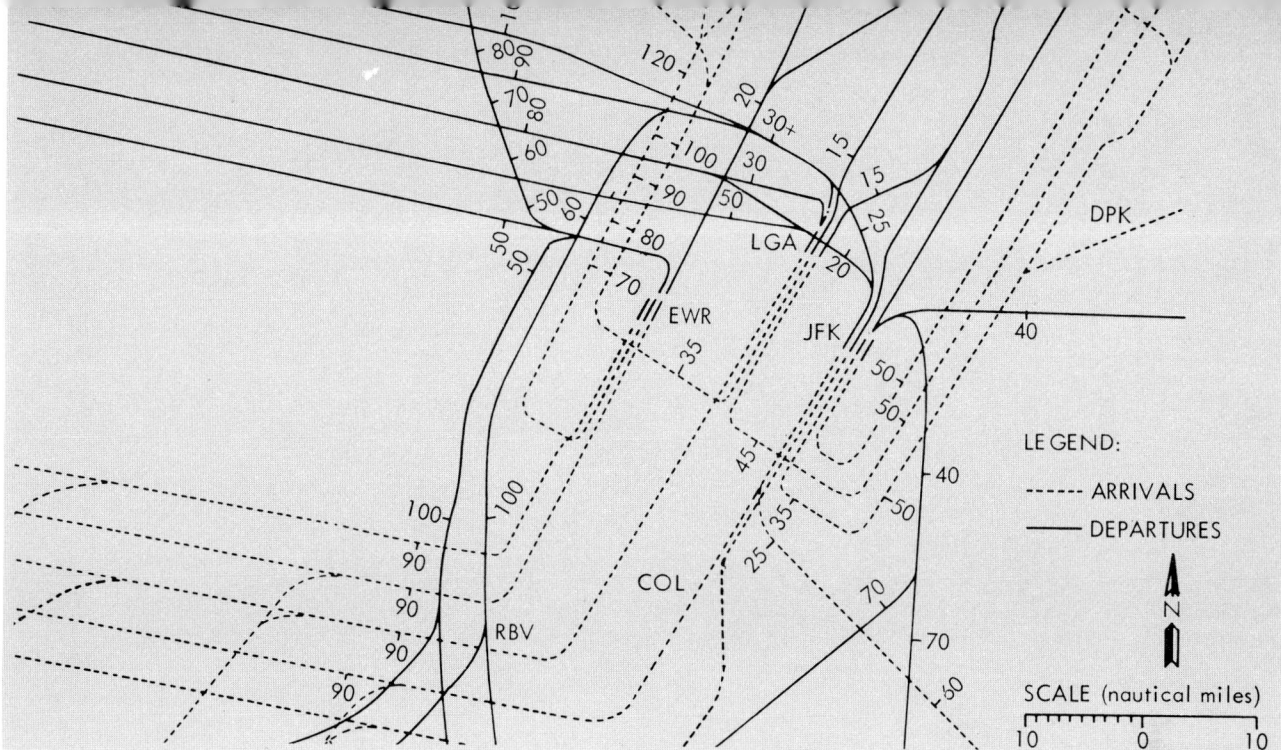

86 *Multiple arriving/departing* RNAV *route system as structured for the New York area.*

fied path, or that some modification to the standard traffic pattern is needed.

When the pilot of an RNAV-equipped aircraft wishes to make an approach under instrument conditions to an airport or runway not equipped with a local landing aid, he uses the waypoint(s) established by an approved instrument approach procedure. By reference to the navigation instrument in his particular aircraft, he guides the aircraft on the published track toward the end-of-runway waypoint, down to the established minimum descent altitude (MDA). If the pilot does not see the end of the runway when he reaches the MDA, he pulls up following a "missed-approach" procedure and either attempts another approach or proceeds to his alternate airport. RNAV instrument approach procedures are shown on special charts.

Airborne RNAV equipment enables thousands of airports all over the world to accept air traffic during instrument weather conditions without the need to have a costly instrument landing system installed at the airport.

As with airports, benefits accrue from the installation of area navigation equipment in aircraft through several en route applications. One of the most attractive of these benefits is the capability provided by the system to fly direct routes between any departure and destination points — rather than having to follow a circuitous airway structure — by establishing suitably spaced waypoints to delineate the desired route. Routes parallel to and offset any desired distance from a direct RNAV route also may be followed.

Another RNAV advantage is that a pilot may fly a parallel course at a specified distance from the VOR radial which marks an airway. For example, one aircraft might be instructed to "fly five miles to the right" of the same radial. Such procedures provide the ATC System with the capability of multiple tracks between terminal areas, using the established airway system, but with positive lateral separation effected between the aircraft concerned.

An additional benefit which is available to the pilot of an RNAV-equipped aircraft is the capability, when circumventing thunderstorms or other turbulence, to keep the ATC System informed of the exact distance he has deviated from the original track.

With the expanded capacity given to the ATC System by airborne RNAV equipment, it is possible merely by establishing waypoints to delineate multiple tracks of any desired configuration for different purposes. Examples include the establishment of discrete routes for flights at different altitudes, for different classes of traffic (V/STOL's, conventional, SST's), and for departing and arriving traffic at airports. These routes can be shown on charts, appropriately identified, thus greatly simplifying ATC instructions. ATC has the flexibility to readily change track configuration as may be indicated by experience or changing traffic conditions, merely by issuing a new description of tracks, based on specified waypoints.

87 *Inertial navigation system (*INS*) control and display unit (*CDU*). When waypoint altitudes are programmed into the* INS *computer, a desired vertical as well as horizontal profile can be followed.*

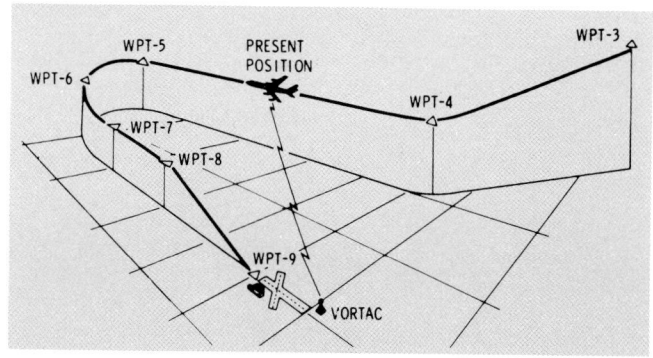

The Air Traffic Control System consists of a complex of ground facilities to assist the controller and pilot in avoiding collisions, yet at the same time permitting expeditious flow of air traffic. Facilities are designed for the various segments of flight from en route operation, transition to an airport area, operation within the airport area, and final approach and landing at the airport.

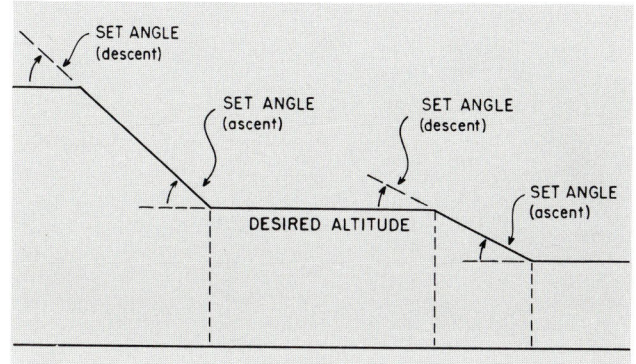

88 In vertical navigation (VNAV), the pilot can arrive at a desired altitude by programming his 3-D RNAV computer with the appropriate ascent or descent angle in relation to 2-D waypoints.

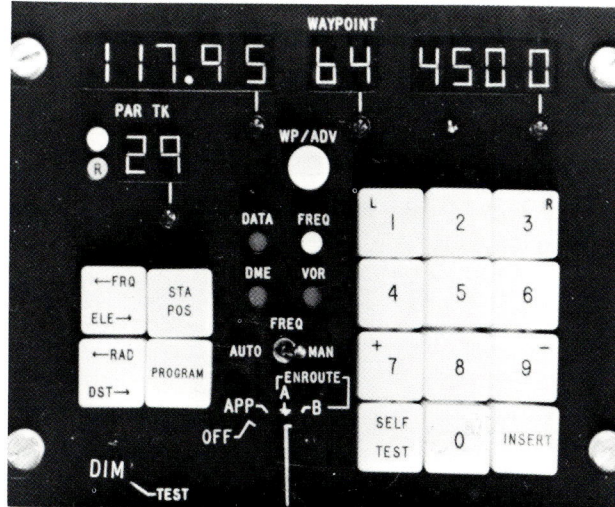

89 Multipurpose CDU can be operated manually or automatically to insert desired 2-D or 3-D flight parameters into the RNAV computer system.

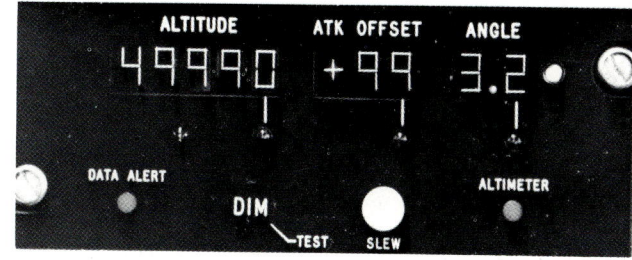

90 VNAV display unit in 3-D RNAV system shows altitude and angle information as programmed by CDU.

Self-contained RNAV

The two basic self-contained RNAV systems used in the ATC System are inertial navigation systems (INS) and Doppler.

Self-contained RNAV equipment employs the latitude/longitude coordinates of the same waypoints used by ground-station-referenced RNAV equipment. Thus, when RNAV-equipped aircraft fly over land areas, the routes, tracks, and waypoints are common. When flying over oceans and other areas where ground navigation signals are not available, the self-contained systems have the advantage of providing the pilot with uninterrupted navigation guidance based on computer memory (dead reckoning) from the last known position.

For accurate area navigation within an active ATC environment, self-contained systems require some form of updating. When over land areas where VOR/DME ground stations are available, the rho/theta position of the aircraft may be used to update the navigation computer system. If two VOR/DME ground stations can be received at the same time, DME/DME or "rho/rho" updating may be more accurate. Other land-based navigation facilities which may be used for updating include Omega and LORAN.

Vertical Navigation

The great amount of flexibility in selecting optimum routes, which is inherent in the 2-D area navigation systems, is considerably enhanced by the addition of equipment which permits the positioning of waypoints at selected altitudes. In effect, waypoints then become "points in space" through which aircraft may be directed to fly at precise altitudes in accordance with ATC instructions. Altitude waypoints may be included in the computer programming described earlier. With vertical navigation, or 3-D RNAV capability, the pilot operates his aircraft in climb or descent so as to reach a specified point in space at a specified altitude. This point in space is referenced to the waypoint being used for horizontal (2-D) navigation. The vertical computer may be programmed so that the aircraft reaches the desired altitude exactly when over the waypoint, or when at some predetermined distance either before reaching or after passing that waypoint. The pilot may select the climb or descent gradient or angle which will be followed by the aircraft in reaching a desired altitude. Standard instrumentation gives "fly up" or "fly down" commands to stay on the selected climb/descent angle in the same manner as when flying an ILS glide slope.

The vertical navigation capability of 3-D RNAV provides the pilot and the ATC System with a high degree of flexibility. For example, three-dimensional paths in space can be established in the form of "tubes" which, in a terminal area, lead the pilot most expeditiously to the runway threshold with the appropriate altitude changes being an integral part of the tube. The only ATC instruction required for this operation is for the controller to identify the selected approach tube in the form of an RNAV STAR. Similarly, on departure the pilot follows another tube in the form of the RNAV SID specified by the controller. No vectoring or altitude-change instructions are required. The pilot does the navigating in three dimensions. The controller monitors, stepping in only as may be needed.

Computer systems for vertical navigation are designed for use within the United States ATC System only if mated with an approved horizontal (2-D) area navigation system. In order to achieve desired accuracy in vertical navigation, "slant-range correction" is necessary. This corrects for the difference in measuring the longer slant-range distance from the aircraft to a ground-reference navigation station and the shorter actual distance on the surface from the station to a point directly under the aircraft. The shorter distance is that which is desired in order to achieve accurate vertical separation between aircraft when using VNAV. Obviously, the higher the aircraft, the greater the slant-range error and the greater the correction needed in vertical position determination.

Time-referenced Navigation

Time-referenced navigation introduces a fourth dimension to area navigation technology. When time functions are coupled with a 3-D RNAV system, the result is 4-D RNAV.

As in the case with vertical guidance, time functions

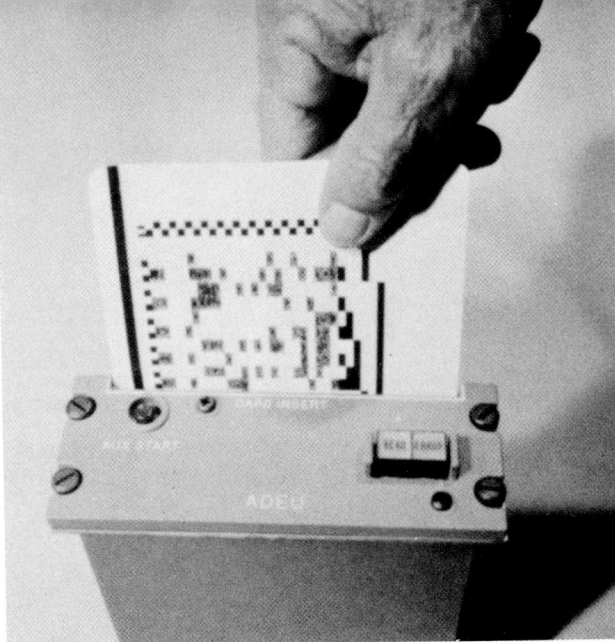

91 Automatic data entry unit (ADEU) uses optical reader techniques. One side of card contains RNAV flight data for pilot reference (routes, SID's, STAR's, IAP's), and on the other side the corresponding symbology for ADEU interpretation and automatic transmission to CDU.

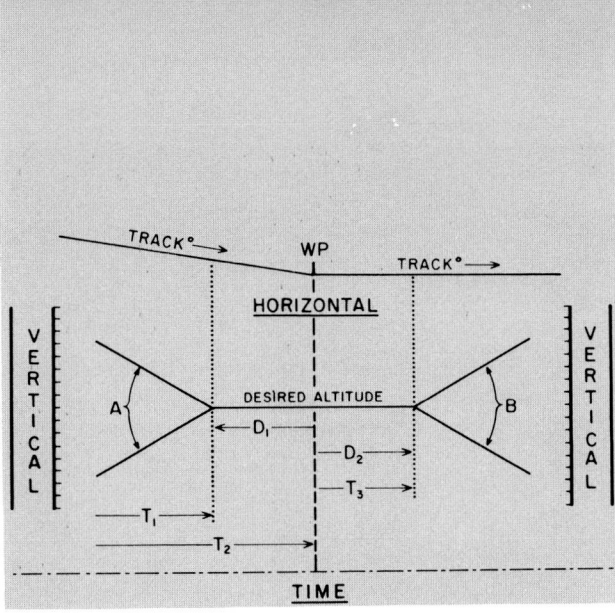

92 In 4-D RNAV, desired time is achieved in relation to distance to or from a waypoint in the same manner as desired altitude is achieved.

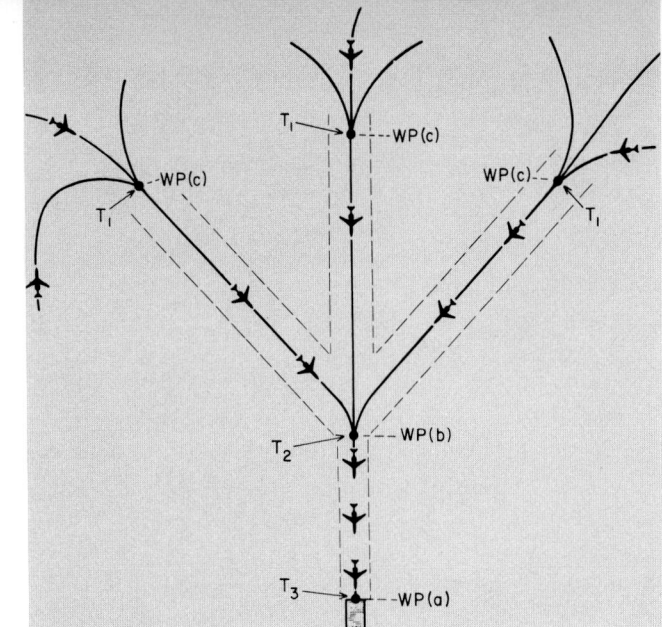

93 Approach control in a terminal area supplies computer-derived metering and spacing commands to provide optimum runway acceptance rate; pilot complies by reference to airborne 4-D RNAV equipment.

94 Pictorial display using a moving map; the aircraft's position is indicated by a moving stylus or "bug."

can be controlled with the same waypoint being used in the horizontal section of the system at any given moment. The principal time function in 4-D RNAV is to give the pilot the ability to arrive precisely at a particular point in space at a desired time. In general, the ATC ground system will establish such desired time (using ground computers as necessary) for purposes of interlacing merging traffic, spacing crossing traffic, metering traffic coming into terminal areas, and sequencing landing traffic.

The pilot, by reference to appropriate cockpit instrumentation is able to adjust power as necessary to make good the desired time, taking any delays involved at higher altitudes and thus minimizing fuel consumption. Conversely, if a desired time as requested by ATC could not be made good, the pilot by reference to his cockpit instrumentation is able to ascertain this in due course and request a time adjustment from ATC. Additional time functions which may be included are: time to go; time of arrival; and ground speed.

The various applications of 3-D RNAV obviously are enhanced by providing the pilot with the capability of arriving at a particular point in space, at a particular altitude, and at a particular *time*.

In the last segment of a flight, that is, the final approach to landing, ATC will advise the pilot of the exact time he should arrive at the runway threshold in order to achieve a maximum runway acceptance rate.

Pictorial Displays

To supplement or augment basic navigation instrumentation (cross-pointer designs) available for use with RNAV systems, certain types of cockpit pictorial displays are available. Generally, these are used with the more sophisticated area navigation equipment installed in the larger aircraft.

A pictorial display assists the pilot with a continuous visual orientation in relation to his navigational environment and flight pattern in real time. He can "see" how he is progressing over the surface or on RNAV routings by reference to background information portrayed on the display in a moving fashion correlated with aircraft position.

The basic types of pictorial displays use either a moving map, a stored film system, or a cathode ray tube (CRT). The CRT offers the most flexibility as any desired information can be stored, electronically generated, and displayed. The CRT display also offers applications for displaying automatically ATC instructions and traffic information.

System-Accuracy Criteria

In the United States, comprehensive criteria are established—and revised from time to time as appropriate—governing design and accuracy requirements for area navigation equipment to be used in the national airspace system. Actions along similar lines also are being developed on a continuing basis within ICAO and other countries having advanced air-transportation systems.

95 Cathode ray tube (CRT) electronically generates real time pictorial display showing aircraft position in relation to navigation routes and facilities; also, auxiliary CRT's show selected flight planning information.

96 Special RNAV simplified charts can be used to follow the most direct VFR routing between departure and destination points; also, to facilitate avoidance of restricted areas, positive control areas, and other airspace in which the VFR pilot may not wish to enter.

RNAV equipment falls into two broad categories: that which can be used only in VFR operations; and that which can be used with IFR operations.

RNAV equipment which meets only VFR standards, as a consequence, is modestly priced. But it can play a significant role. It can help the VFR pilot find airports in inclement, but not IFR weather. It can help him to stay within designated "VFR freeways" and outside of high-density, IFR-controlled airspace, thus contributing to the overall collision-avoidance objective of the ATC system.

In the case of IFR standards, RNAV equipment is in two fundamental classifications: equipment approved for RNAV IFR operations en route and in terminal areas, but not for RNAV instrument approaches; and equipment approved for RNAV IFR operation in all environments without restriction. Accuracy is the basis on which RNAV-equipment approvals and certifications are determined.

In determining accuracy of a particular RNAV system (whether 2-D or 3-D), an "error budget" is analyzed consisting of the following.

- Errors introduced by the ground navigation inputs used (e.g., VOR/DME, TACAN, Omega, and LORAN).

- Errors introduced by the corresponding airborne receiving equipment.

- Errors introduced by the computer (e.g., VOR/DME, INS, Doppler, Omega, LORAN).

- Errors introduced by the pilot (involves resolution of

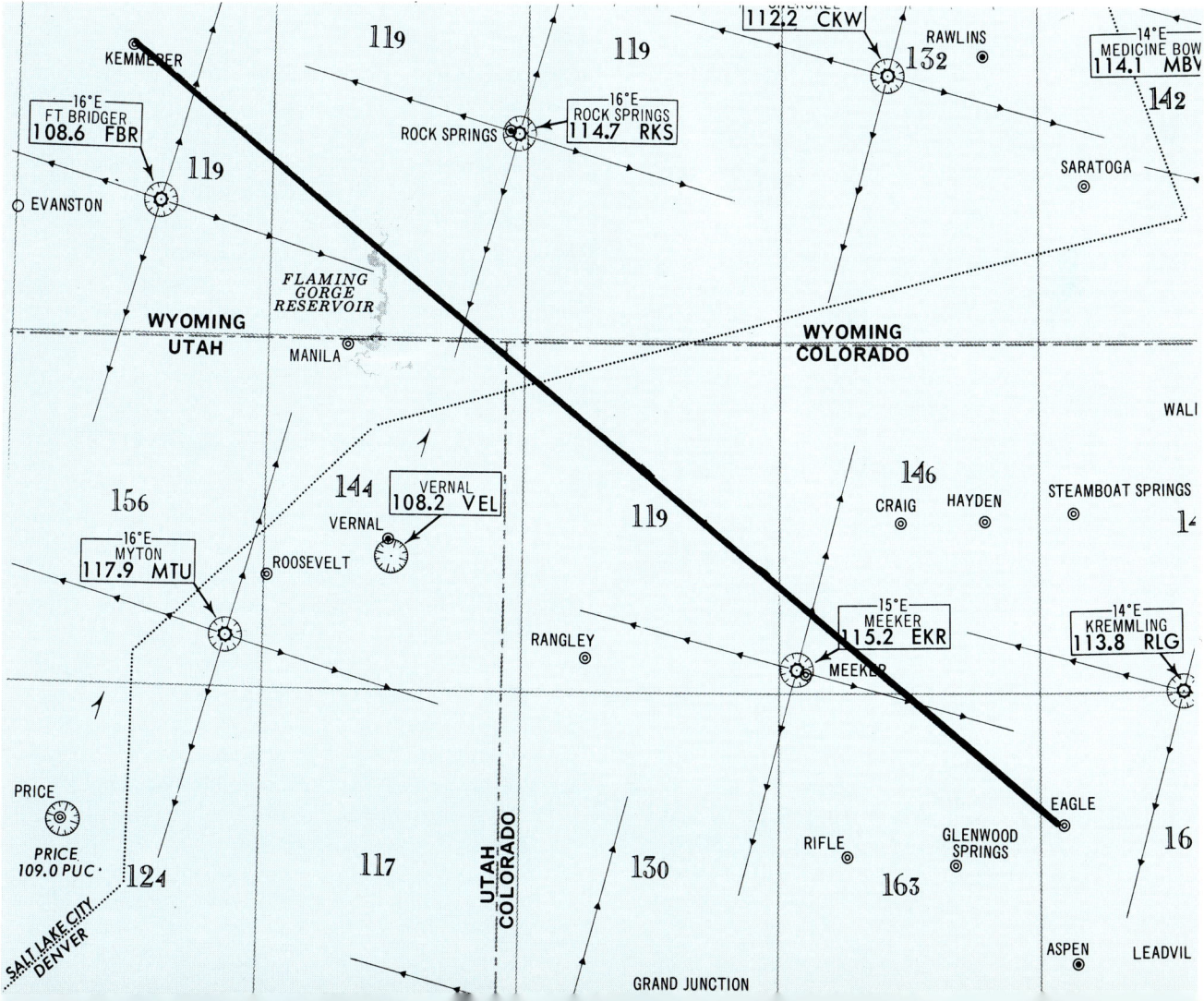

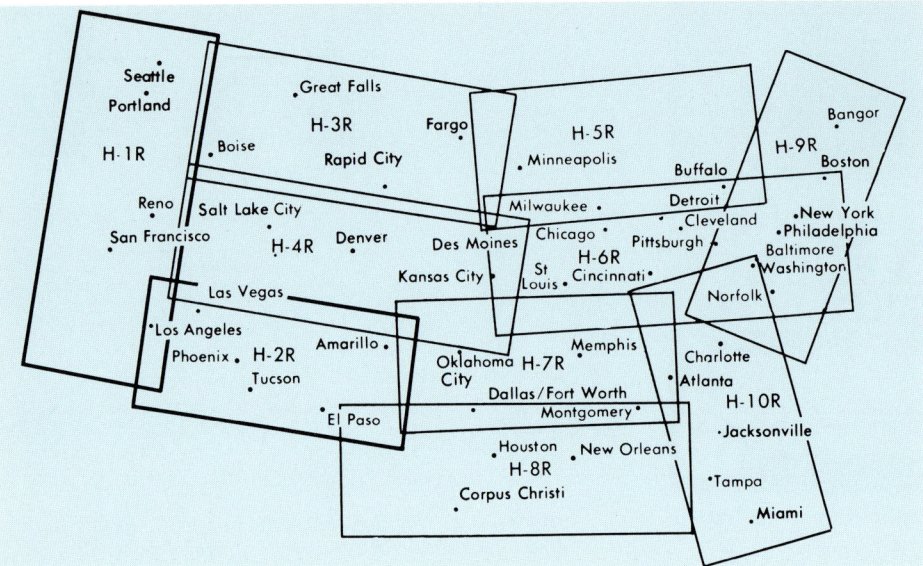

97 Pilot selects appropriate RNAV charts to use in filing an RNAV IFR flight plan with ATC.

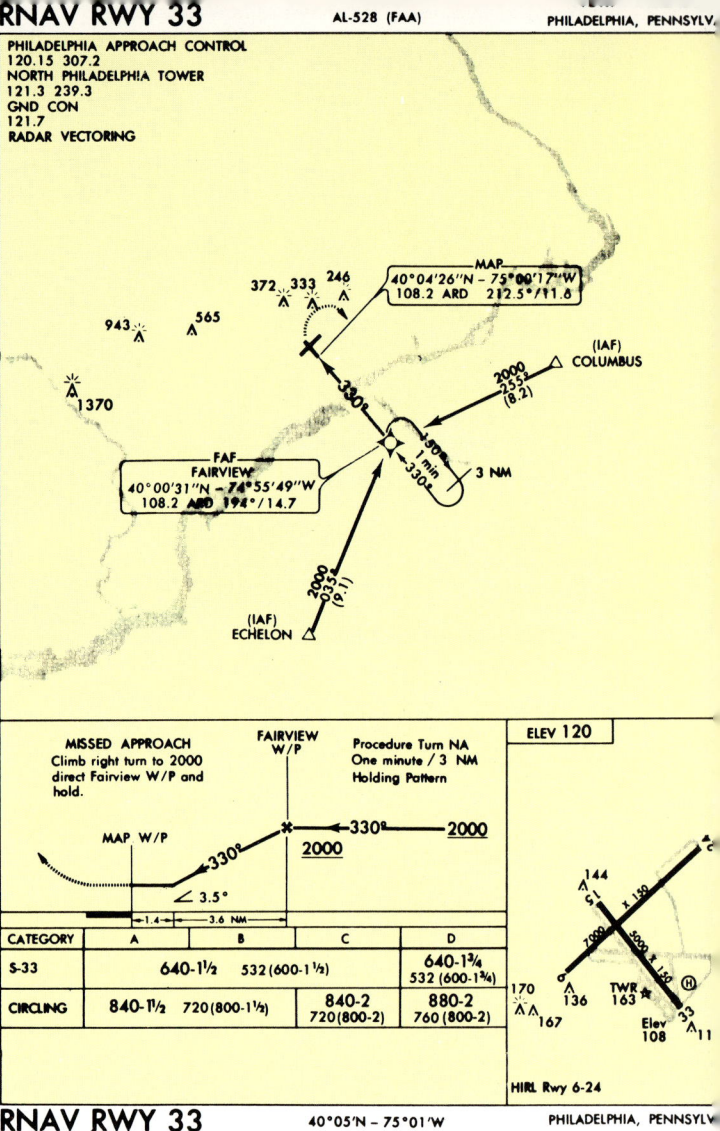

98 In an RNAV instrument approach procedure: above, waypoints define the initial approach fix—IAF (coordinates are shown on en route RNAV chart), final approach fix (FAF), and end of runway—"missed approach" point (MAP or MAWP); and below, RNAV IAP using a scanning beam MLS.

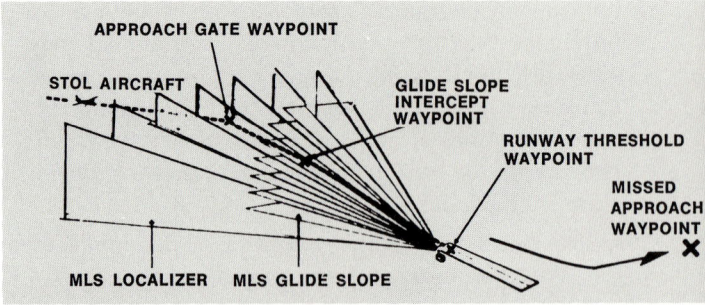

RNAV instrumentation—cross-pointers or pictorial display—and interpretive feasibility).

• Altimetry errors (VNAV only).

Based on the assumption that the variable errors from the different sources normally are distributed and independent, they may be combined in RSS (root-sum-square) fashion. Thus, the deviations obtained from the various error-contributing sources may be combined geometrically rather than arithmetically by taking the square root of the sum of their squares. Calculations made in this manner are required to provide at least 95 percent probability in accuracy of the final accuracy-measurement result.

Other factors affecting approval of an RNAV system for IFR operations within the ATC System involve design characteristics including meeting certain environmental tests and reliability under various operating conditions. Accuracy criteria for IFR certification of RNAV equipment become more stringent as the "state of the art" improvements take place. As a minimum, the accuracy of airborne RNAV equipment equals that of the controller ground-based surveillance radar. In many instances, accuracy is superior.

RNAV Flight Plan

Once an RNAV airborne system is approved for IFR operation (in the United States by the FAA), the controller does not need to be concerned as to the make, model, or manufacturer, since uniform standards have been applied in each case.

Only the pilots of aircraft equipped with an approved IFR RNAV system on board may identify this capability when filing an RNAV IFR flight plan. A special code included in the flight plan indicates this capability. The flight plan may include only one or more RNAV routes or a combination of RNAV routes and VOR airways. The RNAV routes are identified by the suffix "R," preceded by the route number and "J" indicating high altitude (generally above 18,000 feet) or "V" indicating low altitude (below 18,000 feet).

RNAV Instrument-Approach Procedures

In filing an RNAV flight plan, RNAV IAP's are identified by "RNAV" and the runway to which a straight-in (standard) approach is to be made, or "RNAV" followed by reference to a particular circling-approach procedure to be carried out at the airport of destination. RNAV IAP's follow the same concept as the basic instrument-approach procedures. The principal difference, however, is that the locations of the IAF and FAF are defined by RNAV waypoints rather than conventional methods such as VORTAC radial intersections. Another waypoint is located at the end of the runway, called the "missed-approach waypoint" (MAWP), which is used by the RNAV computer to provide the pilot track guidance and distance to runway end (2-D RNAV), and additionally a selectable glide slope (3-D RNAV). Landing minimums at a particular airport depend on such factors as distance from a ground-based navigation aid, obstruction clearance, and whether the approach is based on the use of 2-D or 3-D airborne RNAV equipment.

CHAPTER IX

AIRCRAFT

99 The changing times in general aviation instrument panels; upper photo about 1940; lower, the 1970s. General aviation area navigation instrumentation includes (1) waypoint coordinate selector; (2) track selector and display.

Aircraft are to the control of air traffic as ground vehicles are to the control of surface traffic. Both types of traffic control must cope with vehicles of different sizes and speeds with peak traffic surges, and with weather conditions which may slow down otherwise normal traffic flow. Moreover, both systems have something else in common—preventing collisions and expediting vehicular movement.

In air transportation, however, the aircraft—or air vehicle—moves in three dimensions. This gives an advantage to the Air Traffic Control System over ground traffic control because of the added flexibility provided by the third dimension—height. On the other hand, the third dimension makes air traffic control more complicated in principle than surface traffic control because of problems in having aircraft accurately follow specified three-dimensional "highways," "overpasses," and "underpasses" in the sky.

In another area of difference between surface and air traffic control, surface vehicles can slow down or stop as traffic conditions indicate. Conventional-type aircraft cannot. They must keep moving at minimum speeds, which in most cases exceed the maximum speed of ground vehicles. But this picture changes with aircraft which can land and takeoff vertically or at slow speeds. In this case, the comparison between ground and air traffic control again becomes closer.

An area of broad similarity in the control of ground vehicular traffic and the control of air traffic is that both have different classes of users. They each have varying needs for different types of ground or air vehicles: from compact automobiles/small single-engine aircraft to huge buses and trailer trucks/jumbo jet passenger and freight airliners. Yet, all must live together in their respective environments—surface and air.

AIRCRAFT CATEGORIES

Aircraft comprise one of the most important elements in the ATC System. Their characteristics directly affect the efficiency of other elements of the system, such as traffic management, airspace, and airport utilization. As aircraft designs, configurations, and performance characteristics change, the ATC System must be modified accordingly.

100 Long-range, wide-bodied airliner.

101 Medium-range, wide-bodied airliner.

102 Medium-range, standard-configuration airliner.

Or, looking at it another way, the ATC System must be capable of accommodating all classes and categories of aircraft to permit them to take maximum advantage of their particular role in the overall air transportation system.

From an *operational* point of view, aircraft are divided into two categories: those qualified to be flown under instrument flight rules (IFR), and those restricted to flight under visual flight rules (VFR). An aircraft of a given make or model may be certified for either IFR or VFR, depending upon its instruments, navigation, and radio equipment. Other makes or models may be restricted to VFR because of airworthiness considerations. Air Traffic Control does not enter into this question when a pilot files a flight plan, nor whether the pilot is qualified to carry out the proposed flight. It is the direct responsibility of the pilot to fly a particular aircraft only under operating conditions for which he and the aircraft are authorized.

From a *performance* point of view, aircraft fall into the following categories.

- CTOL—conventional takeoff and landing aircraft.
- V/STOL—vertical and/or short takeoff and landing aircraft.
- SST/HST—supersonic/hypersonic transport aircraft.

CTOL AIRCRAFT

Aircraft in this category range from the subsonic jet transports to the small fixed-wing private aircraft. Speeds of the CTOL aircraft range from 600 to 80 knots cruising, and from 160 to 30 knots during landing and takeoff. CTOLs are classified from an air traffic control viewpoint as "high performance" and "nonhigh performance." High-performance aircraft include all turbine-powered aircraft—airline, general aviation, and military—and those propeller-driven aircraft which can operate at the same speeds in a terminal area (150-250 knots) as turbine-powered aircraft.

It is obvious that mixing aircraft of widely differing performance characteristics could lead to conflict or mid-air collisions. This is particularly true in terminal areas and at airports of medium- to high-density traffic volume. As a consequence, a general objective of ATC in airspace structuring is to segregate low-performance from high-performance aircraft to the greatest extent possible. This is not to deny any aircraft operator use of his vehicle, but rather to provide separate or "discrete" routings and runways wherever needed for safety.

The introduction of wide-bodied or jumbo jets during the early 1970s, mainly by the trunk and regional air carriers, provided a significant increase in air-transportation capacity without a corresponding increase in the number of aircraft. Thus, from the standpoint of holding down the volume of air-carrier aircraft that need to be controlled by the ATC System, the introduction of these aircraft was a success.

But, these aircraft introduced a new equation in the separation of aircraft by ATC—the effects of wake turbulence. Wake turbulence, caused by rapidly rotating air currents trailing the aircraft's wing tips, cannot be seen by the pilot of a following aircraft. When a large aircraft lands, such as the B-747, DC-10 and L-1011, this turbulence can lay on the threshold of the landing runway for an indeterminate time, depending on wind conditions. Wake turbulence occurs similarly on takeoff of a large aircraft. Wake turbulence is not formed after the aircraft is on the runway or before rotation on takeoff.

The effect on the ATC System of wake turbulence—which can put a following aircraft, even a medium-size jet, out of control—is to provide greatly increased separation between a large aircraft and a following aircraft, either on landing or takeoff. Subject to aerodynamic modifications to large aircraft which reduce or eliminate wake turbulence, or availability of accurate detection means, this phenomenon is a serious limiting factor in the ATC System capacity wherever the large aircraft fly.

In addition to the relatively small number of high performance, and highly productive, CTOL aircraft used by the air carriers, CTOL's of all categories in constantly growing numbers are used by general aviation. Most CTOL's used in personal and instructional flying are propeller driven with one or two engines, and those used in business and commercial flying generally are multi-engine turbine-powered aircraft.

Military CTOL's include high-performance transport aircraft of a type similar to the large air-carrier jets, as well

103 Short-range airliner.

104 Military transport aircraft.

106 First utility helicopter to be fully certificated in the United States for instrument flying (early 1970s).

as various configurations of fighter bomber and training aircraft.

The CTOL aircraft will comprise the bulk of the air traffic with which the ATC System will contend during the 1970s and 1980s. The variations in missions, capabilities, and performance of the different types of aircraft within this category require great flexibility and accommodation in ATC system design and operation.

V/STOL AIRCRAFT

"V/STOL" means those types of aircraft that are capable of making a vertical takeoff and landing (VTOL) and/or a short takeoff and landing (STOL). While a STOL aircraft does not have vertical takeoff and landing capabilities, all VTOL's have STOL capability to some extent. To qualify as a STOL, generally speaking, an aircraft must be capable of taking off or landing over a 50-foot obstacle at sea level at its maximum takeoff or landing weight in a distance of about 2,000 feet. A variation of this concept is an RTOL (reduced takeoff and landing) aircraft which can operate into and out of runways about 4,000 feet in length. A true VTOL requires no distance for a landing or takeoff roll, although in some designs a short takeoff run increases payloads.

The ever-increasing demands made by CTOL aircraft on airport capacity has created many extremely severe problems. Included are the constantly increasing number and duration of air traffic delays, lack of large runway expan-

105 Typical general aviation aircraft used in corporate or executive flying, business flying, and personal flying.

sion capability at existing airports, and lack of real estate to construct new conventional airports to meet growing urban and intraurban air-transportation needs. New approaches to aircraft design and performance offer the opportunity to increase capacity and productivity of airports as well as the ATC System in general.

Integration of V/STOL aircraft into the air-transportation environment provides a tremendous potential for relieving congestion problems at airports, as well as making possible the development of new air services. To accomplish this objective, however, some major changes in ATC philosophy and techniques are required involving the use of this type of aircraft. The capability of V/STOL's to utilize short runways or merely small "pads" or strips makes it possible to establish separate and distinct (discrete) landing and takeoff paths. This capability, in turn, permits segregating V/STOL and CTOL traffic in order to achieve maximum use of airspace and airports by both categories of aircraft. V/STOL's open up a new dimension in air transportation.

Helicopters

This member of the VTOL family includes any type of airborne vehicle which derives its principal vertical lift capability through the use of some form of rotary wing (i.e., blades or other airfoil) mounted on a shaft and caused to rotate in the horizontal plane of the aircraft by power-driven means.

107 *Corporate or executive passenger helicopter.*

108 *Helicopter "crane."*

109 *Military heavy-lift helicopter, the "super Jolly Green Giant."*

110 *Commercial heavy-lift helicopter.*

111 *Personal use helicopter.*

Helicopters were "born" in 1940, using piston-powered engines and with many "bugs." They were very slow in development and generally were considered aeronautical freaks. The Korean War, however, gave their development an impetus and demonstrated their practical utility. Further improvements in vibration, stress, and helicopter aerodynamics followed, and these first generation helicopters completed their cycle in the late 1950s. The second generation helicopters came into being at that time, and helicopter production advanced all over the world with many aerodynamic refinements, such as introduction of turbine power plants in multiple configurations, and increased payload capability. The third generation of helicopters commenced in the late 1960s with much greater speed, impressive weight-lifting capacity, relatively simple design, lower costs, and safer operation.

The third generation helicopters continue the original "pure" configuration as well as moving into the "compound" and "rotor/wing" concepts. The pure helicopter is one in which all the engine power is used to drive the lifting rotor or rotors, which also supply the thrust for forward flight. The compound helicopter, in addition to the rotors, has a fixed wing—like a conventional airplane—plus a supplemental thrusting power plant (jet or propeller driven) to give more forward speed. In the rotor/wing design, the rotor/wing acts as an ordinary helicopter rotor during ascent and descent with the exhaust of a turbojet power plant driving the rotor through tip jets; after reaching a certain speed, the rotor becomes stationary and the vehicle cruises as a conventional fixed-wing, jet-propelled aircraft.

Pure helicopters in the third generation are capable of speeds up to 225-250 mph. Compound helicopters up to 350-400 mph and rotor/wing helicopters up to 500 mph will follow. Another major developmental area in the third-generation helicopter is the increase in payload. Payloads of 5-10 tons in military application have been an operational reality as evidenced during the Vietnam War, as well as 25-ton-payload crane helicopters.

Folded-Rotor Compound Helicopter

The main rotor is a three-bladed rigid rotor type with blade folding features. The rotor supplies lift during hover and transition, but for conventional flight, the rotor is stopped and folded so that all blades trail aft from the pylon. Lift in conventional flight is supplied entirely by the wing. Turboshaft engines drive either the rotors or the cruise propellers, and permit the aircraft to attain considerably greater speeds than the pure helicopter.

Nonhelicopter VTOL

This category comprises, for the most part, types of aircraft which attain vertical flight by means of downward thrust using propellers or pure jet engines. Most of these VTOL aircraft are designed for high-speed operation when in horizontal flight. Several basic designs are included in this family of VTOL's.

112 *Tilt-wing* VTOL.

113 *Tilt-ducted propeller* VTOL.

114 *Jet lift-cruise* VTOL.

115 *All-purpose utility* STOL.

116 *Medium-capacity* STOL *and large* (AMST) *military* STOL.

Tilt-Wing VTOL

The tilt-wing feature and ample turbine power provide vertical-lift capability using propeller thrust alone—the wing may tilt to 100° from horizontal plane of aircraft. After takeoff, the wings gradually are changed to normal attack angle to provide for conventional horizontal flight.

Tilt-ducted Propeller VTOL

Ducted propellers driven by turbine engines are rotated to vertical thrust position for takeoff and landing, while the wing remains fixed. For forward flight, the ducted propellers are transitioned gradually to horizontal thrust position.

Jet Lift-Engine VTOL

A number of small direct-lift engines are mounted in pods on the wings and perhaps the fuselage. They are used only during takeoff and landing, the cruising thrust power for the aircraft being provided by conventional turbojet or turbofan engines.

Jet Lift-Cruise VTOL

Lift is derived from the use of power generated by jet engines which also power the aircraft in cruise flight as a turbojet or turbofan thrust system. For vertical takeoff and landing, the hot gases of the jet engines may be ducted to drive wing fans (plus perhaps nose and/or fuselage fans); or wing-mounted nozzles or a jet-flap (divertor) may be used to deflect the hot gases downward for lifting capability.

STOL Aircraft

In most cases, this is essentially a conventional aircraft, with its STOL characteristics being attained by low wing-loading, full-span propellers, slip-stream coverage (outboard propellers may have differential pitch in four-engine types), and extensive use of high-lift aerodynamic devices such as flaps and slats. Various designs are powered by turbine turboprop/turboshaft engines, but some models are powered by piston engines. Other designs use pure jet propulsion.

V/STOL POTENTIAL

In commercial aviation, the V-STOL's forseeable future basically is as a short-haul transport. Although ranges up to about 750 miles have been considered in different designs, the 500-mile and under range generally appears to be optimum.

By far, the great majority of people throughout the world travel by automobile on trips under 200 miles and within metropolitan areas. Yet, is the automobile capable of providing efficient urban mass transportation? The answer is indicated by considering some statistics derived from experience in the major population centers of the United States. At peak periods, automobile traffic moves at speeds averaging 6 to 11 miles per hour. A man's normal walking pace is 4 mph. An average automobile carries 1.3 persons; thus, about 80 autos are needed to carry 100 passengers. Eighty autos require about one-third of a

117 *Concept of mobile pod for use with flying helicopter crane.*

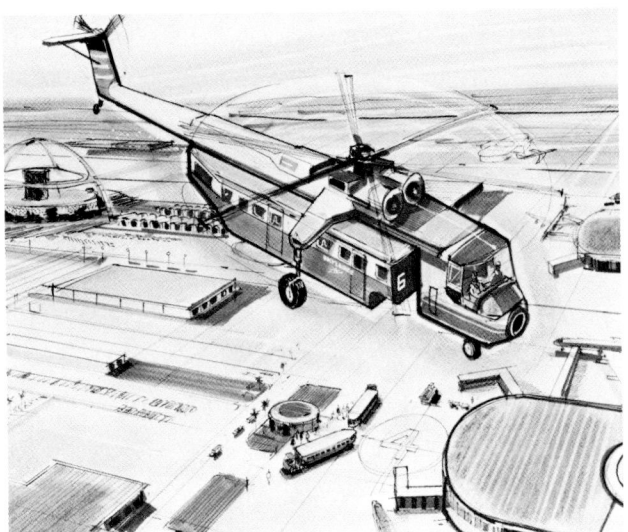
118 *Mobile pod attached to flying crane.*

119 STOL *city-center operation.*

mile of "freeway." One mile of freeway removes 120 acres from tax rolls.

It seems obvious that automobile transportation cannot continue to expand indefinitely in high-density-population areas to meet increasing short-haul transport requirements. Other solutions must be found, among which the most significant are high-speed surface trains, subways, and aircraft—notably the V/STOL. Massive introduction of V/STOL aircraft into short-haul air services will impact the ATC System in many respects. Solutions to anticipated problem areas need to be developed well in advance so that the ATC System will not be a factor limiting future V/STOL short-haul air-transportation progress.

Airport-to-Airport Service

This type of service has been provided by helicopters for a number of years, both in the United States and Europe. Its effectiveness for V/STOL's is most apparent when the airports concerned are within a reasonably short distance of each other and the service is not suitable or economic for CTOL aircraft. Experience has proven that the operation of V/STOL aircraft from the same runways and using the same approach and departure paths as CTOL aircraft seriously derogates the effectiveness of the V/STOL aircraft. Because of the short duration of their flights, delays can drastically affect their operation, including direct operating costs (DOC) and reliability. For example, on a short-haul flight of 30 minutes, a 15-minute air traffic delay will increase DOC by *50 percent*; if the delay is 30 minutes, the DOC increase is *100 percent*. These same delays on the longer range trunk and regional flights have minimal percentage impact on their DOC. V/STOL aircraft thus need to be able to operate within the ATC environment independently of CTOL aircraft.

Airport-to-City Service

In this application of V/STOL's, the service pattern involves the collection and distribution of passengers between the large conventional airports and convenient city-center "VTOLports" or "STOLports" within the associated metropolitan area. Generally speaking, these flights would involve distances under 50 miles; however, with the advent of "regional" airports serving a combination of nearby metropolitan areas, the flight distances could extend over a radius up to perhaps 100 miles.

Aside from the use of passenger-carrying V/STOL aircraft, one different approach may involve the use of a helicopter "flying crane." Under this concept, a ground vehicle or "pod" is picked up after loading, either at an airport or at a downtown location, by a flying crane and deposited at its appropriate destination. If the unit picked up by the flying crane is a self-propelled vehicle/bus, the bus can gather the passengers in the downtown area and proceed to a central pickup terminal. From there, it can be attached to the crane and be carried to a landing area adjacent to the conventional airport where the bus would then proceed under its own power to distribute passengers directly to the appropriate departing aircraft loading position.

City-to-City Service

In the United States a high percentage of intercity travel is between 40 pairs of leading cities, the cities in each pair being separated by less than 500 miles. The balance of high-density intercity travel is between an additional 10 city pairs which are 500 to 700 miles apart. In Western Europe, the bulk of intercity travel is between about 30 cities, with virtually all interconnecting routes having stage lengths less than 500 miles. Numerous other high-density, intercity routes of 500-mile stage lengths or less exist or will be developed in other parts of the world. All of these short-haul transportation needs fit advantageously into the future V/STOL service capability, provided, of course, that city-center VTOLports and STOLports are developed.

V/STOL service between city centers presents one of the more glamorous applications of air transportation. It can also be one of the more efficient applications. Because of its environment, however, it requires solutions to a great many problems—technical, economical, and ecological. These involve such factors as noise, obstructions, landing site costs, and navigation difficulties caused by high buildings.

120 Large commuter-type STOL.

121 Concept of VTOL commuter service.

Commuter Service

One of the knottiest problems to be solved during the remainder of this century will be to achieve a breakthrough in urban mass transportation in densely populated areas of the world. In the United States, for example, about 50 percent of the population is concentrated in three supermetropolitan corridors—Washington-Boston, Buffalo-Chicago, and San Francisco-Los Angeles. By early 1980, these three corridors plus the Jacksonville-Miami and San Antonio-Dallas-Houston areas will contain about 160 million people—well over 60 percent of the country's then expected total population.

To visualize the possible application of the V/STOL's in helping to solve the megalopolis transportation problem in the United States, a short-haul (air-bus) air transportation system may be postulated in the "northeast corridor" of the United States from Maine on the north through Virginia on the south—a distance of about 500 miles, and extending inland roughly 140-150 miles. Fifty airport locations would effectively serve the 43 million people expected to reside in this corridor by the 1980s. These 50 sites would consist of conventional airports plus a system of VTOLports and STOLports in downtown locations in the larger metropolitan areas. To service this pattern, about 65 V/STOL air-buses capable of carrying 100 or more pasengers each could be put into operation initially. These aircraft, making about 2,000 stops a day, could carry some 50,000 passengers daily.

In V/STOL air-bus or commuter-type of operation such as that postulated, there obviously would be marked peak loads, such as in the mornings when the commuters are going into the metropolitan areas and in the afternoons when they are returning to their points of origin. To obtain maximum utilization of the V/STOL's, it is visualized that they would be used during the slack hours of each 24-hour period to pick up and distribute cargo, express, and perhaps mail within the service area.

V/STOL TRAFFIC VOLUME

The short-haul nature of V/STOL's will result in their making many more landings and takeoffs in any given period of time in commercial operation than conventional airliners. V/STOL airliners may be operated to perform typical flights of around 7-15 minutes' duration in airport-city service; 15-20 minutes in commuter service; and around 45-90 minutes in city-to-city service. Average operations per hour—landings and takeoffs equal one operation each —would be about as follows.

Service	Average distance	Average block speed	Flight time	Ground time	Operations per hour
Airport-City	20 mi.	100 mph	12 min.	8 min.	6
Commuter	50 mi.	180 mph	17 min.	6 min.	5
City-City	300 mi.	300 mph	60 min.	20 min.	1.5

122 Projection of different categories of aircraft by type.

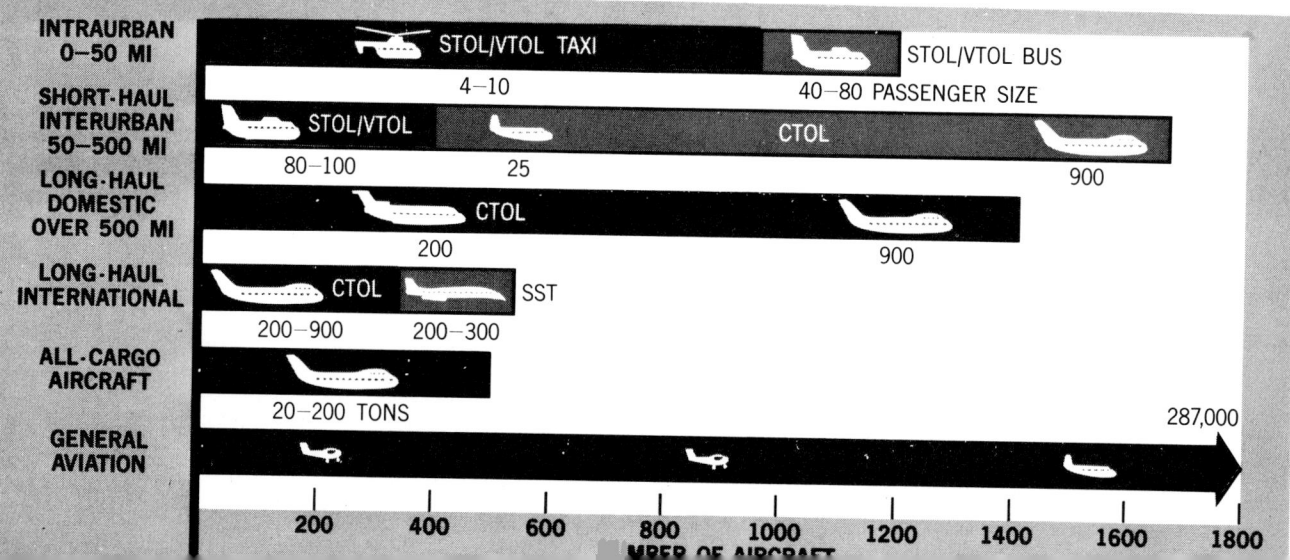

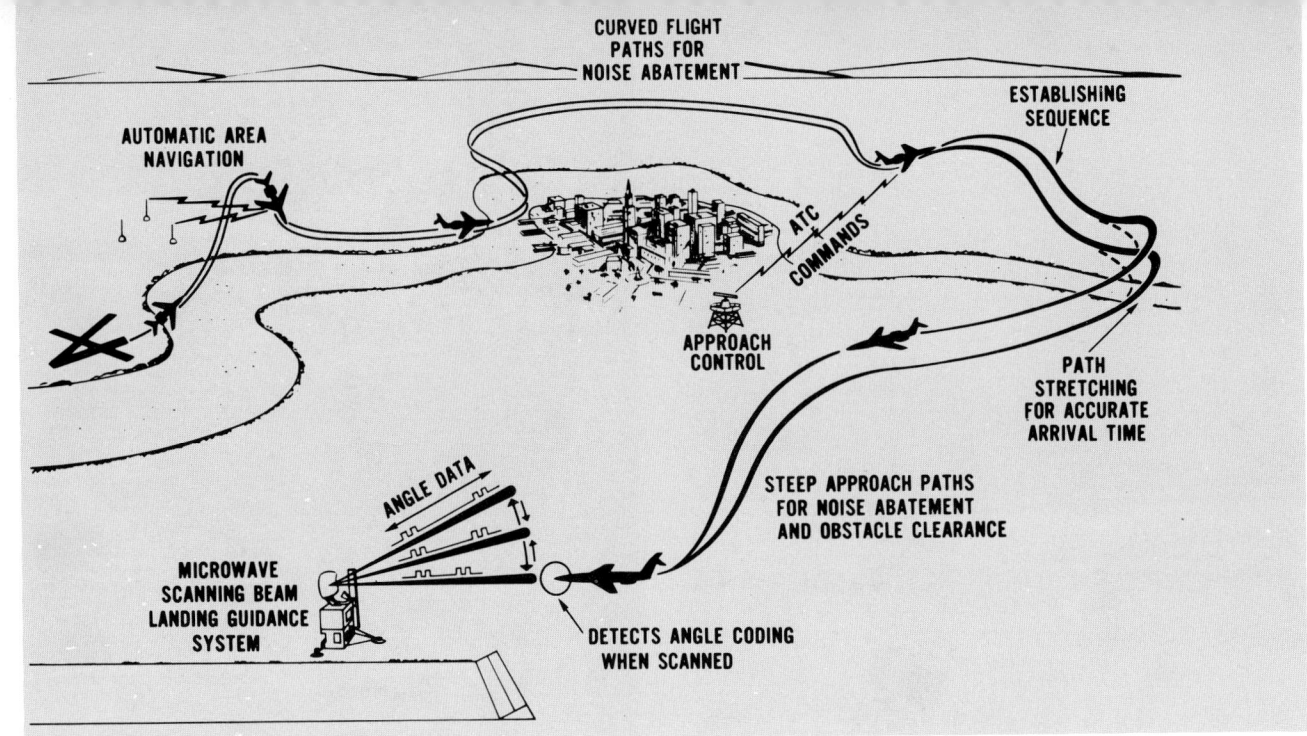

123 V/STOL *characteristics provide distinct advantages over* CTOLS *from the standpoint of* ATC *interface.*

Assuming a more or less equal numerical distribution of the different types of V/STOL's in the above three service patterns, one V/STOL airliner would perform an average of about 3⅓ operations per hour. Given a service utilization of around 15 hours daily, each V/STOL airliner could thus be expected to perform about 50 landings and takeoffs daily. Aircraft used in trunk air-carrier service perform about 6-8 operations daily; aircraft in regional air service about 8-12 operations daily.

In approximate terms, it may be considered that each V/STOL introduced into airline-type service will perform on the average about *five times* as many landing and takeoff operations per day as the average conventional jet airliner. Since landings and takeoffs constitute the most critical phase in the control of air traffic, it is obvious that the advent of V/STOL's in commercial-airline, all-weather service under IFR will tremendously increase the magnitude of the air traffic control problem. For example, 90 V/STOL airliners operating in the northeast corridor of the United States would about *double* the total IFR operations of all CTOL aircraft normally flying within that area.

Developing Area Service

The V/STOL also has a significant role to play in providing short-haul air-transport services which may open up underdeveloped areas of the world where surface transportation is inadequate and the cost of constructing conventional airports is not feasible economically. This is not a new approach to air transportation, as many helicopters and STOL-type aircraft have played an important part in the past in opening inaccessible parts of the globe by airlift. On the other hand, the new generations of helicopters (from the smaller models up to large commercial versions) and the nonhelicopter V/STOL's, with their improved operational and economic performance, can very well play a more significant role in contributing to progress in the developing areas of the world.

General Aviation

Although considerable attention has been focused on the commercial application of V/STOL-type aircraft, their present and future use by general aviation for private and business purposes is not to be overlooked. In this category, most of the V/STOL aircraft used will probably remain in the helicopter category with a capacity of three to seven persons. Large corporations may also become users of the bigger V/STOL's in the same way as they have developed corporate use for the larger conventional turboprop or jet aircraft. The advent of utility helicopters capable of operating under all-weather conditions opens up many new general aviation applications which, in turn, impact the ATC System.

V/STOL OPERATIONAL CONSIDERATIONS
ATC Interface

The capability of the V/STOL to reduce speed rapidly and if necessary fly at slow speeds provides an exceptionally important tool to the ATC System not available with CTOL aircraft. This capability facilitates achieving maximum utilization of the airspace by helping the controller to provide optimum sequencing and spacing of the individual aircraft involved in any particular traffic situation. With the proper application of three-dimensional flight profiles and proper timing-control factors, IFR V/STOLS may be expected to be able to operate effectively and efficiently in IFR conditions following discrete routings en route and to and from vertipads/stolstrips at CTOL airports, as well as to and from segregated VTOLports and STOLports.

In terminal area operations, the various advantages of the V/STOL's flight characteristics—ability to slow down readily, variable approach and climb gradients, and small landing area requirements—introduce many factors which favor IFR V/STOL operation in the ATC environment. The steep approach angles which can be performed by the V/STOL provide a very significant advantage by minimizing the amount of the airspace required for the approach flight path, as contrasted to CTOL aircraft.

Fallouts of this capability include reduced objectionable surface noise-level exposure and the ability to achieve safe obstacle-clearance altitude during the approach. In some

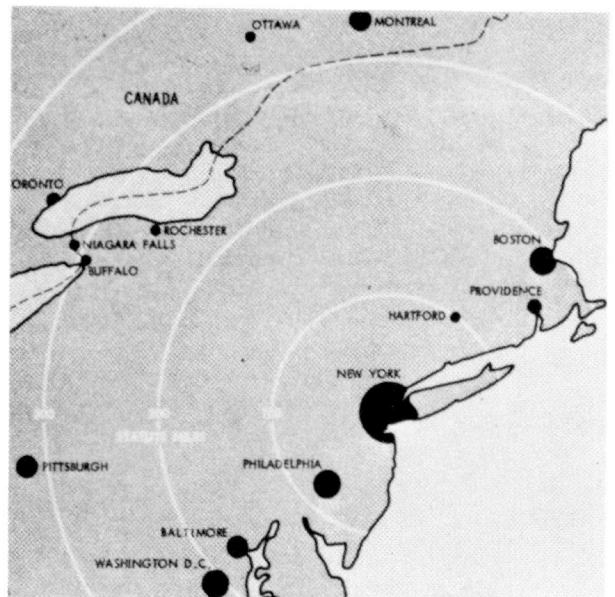 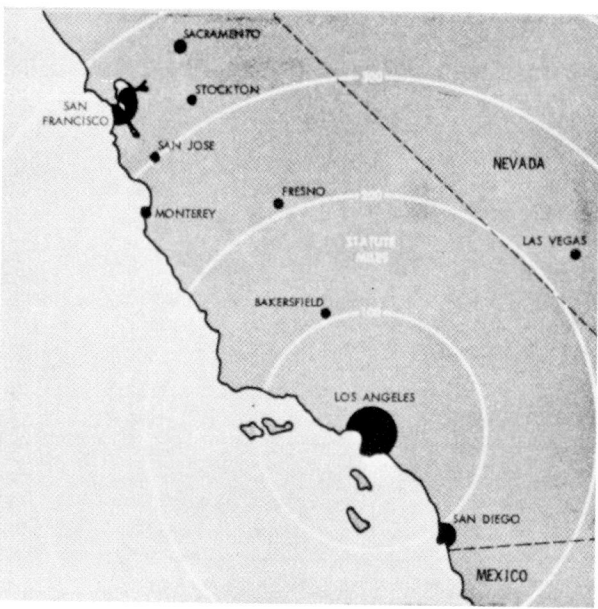 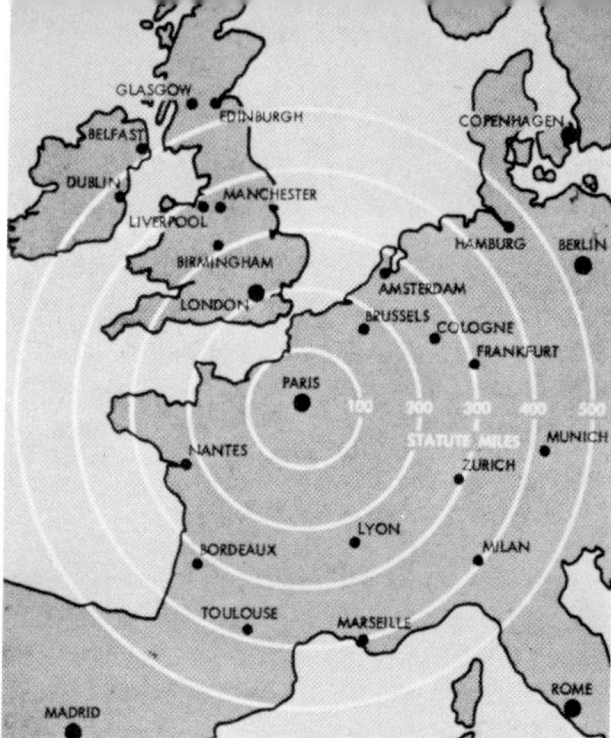

124 Typical densely populated areas within which high volume V/STOL traffic will develop.

instances, "steep approaches" may be the only way that V/STOL's could provide service to a city-center VTOL/STOL-port due to other buildings and structures in the vicinity; also the reduced noise-exposure area may be essential to achieve public acceptance of V/STOL operation. On the other hand, where traffic obstruction and noise considerations aren't significant, the so-called "normal approaches" may be followed.

A two-segment combination of the steep-approach and normal-approach profiles may be found desirable in some instances. In this procedure, a steep initial (first segment) approach would be made to a desired altitude at the desired point in space relative to the touchdown point, with subsequent transition to a normal (second segment) gradient for the final approach to touchdown.

An important part of the equipment of a V/STOL for IFR operation in the ATC environment will be its capacity to navigate with a high degree of accuracy. For arriving IFR V/STOL operations in a terminal area under ATC, the pilot must be able to navigate accurately on specified and identified three-dimensional flight paths, in the same manner as en route, but with higher equipment sensitivity and resolution. The pilot follows a designated initial approach fix (IAF) at the specified initial approach altitude using his 3-D RNAV equipment for three-dimensional flight-profile navigational guidance. Time of arrival at this point may or may not be an air traffic control factor, but if the aircraft is not equipped to permit the pilot to comply with a time assignment from the ground ATC System by reference to airborne equipment (4-D RNAV), the ground system may give speed control instructions to meet desired time parameters.

The final approach may be carried out by reference to a ground-based precision landing system such as a scanning-beam MLS (microwave landing system), or the final approach may be made by the pilot by reference solely to his 3-D airborne area navigation equipment. Also, a "point in space" approach may be made using 3-D RNAV, and if in visual contact with the surface at MDA (minimum descent altitude), final approach and landing may be completed under "special VFR" criteria. In the United States' ATC System, a helicopter or other suitable VTOL may be authorized to fly under special VFR if the pilot is able to maintain visual reference to the surface and the traffic patterns, routes, and reporting or holding fixes. In some instances, the IFR V/STOL pilot may wish to use a conventional ILS for final approach, but this introduces the exposure to being handled as a CTOL aircraft by ATC.

Radar Surveillance

In the past, ground radar surveillance for the control of air traffic in a limited area and with a limited volume has provided an effective tool for the ATC System in organizing traffic flow as well as in preventing collisions between aircraft. It will continue to play an important role in the system. However, with V/STOL's added to other high-density air traffic, it is possible to visualize something in the order of 1,000 aircraft in flight at a given moment within a 100-mile radius of major metropolitan population centers. With this quantity of targets on a radarscope, particularly if each blip is associated with alphanumeric data, interpretation of the radar display and individual direction of each aircraft by controllers would present an impossible problem. In view of the high percentage of low-altitude operation of V/STOL's, the line-of-sight limitation of radar coverage also reduces radar usefulness. A question, therefore, is raised as to whether the concept of air traffic control by ground-interpreted radar surveillance will be feasible in handling V/STOL high-density traffic. This limitation will be even more emphasized as V/STOL's become used in greater volume by general aviation, with the private pilots wishing to use many and varied landing spots.

Air-derived Separation Assurance

The advent of V/STOL high-density traffic in confined altitude strata will lend impetus to the need for some sort of air-derived, separation-assurance system for collision avoidance. This may take the form of a "backup" safety device, or as an integral part of the ATC System. In addition to collision-avoidance performance laterally and horizontally, the V/STOL-collision avoidance or separation-assurance equipment will need to provide coverage during steep climb and descent up to and including vertical

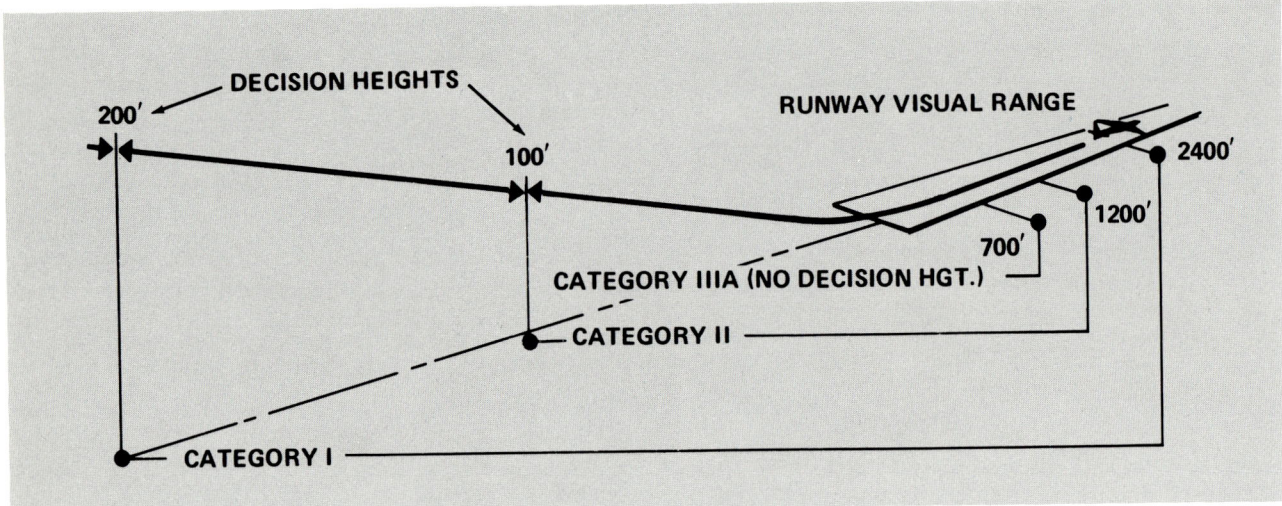

125 Weather categories applicable to v/STOL and CTOL landing operations.

Various types of aircraft permit a diversity in meeting different air transportation requirements. Conventional aircraft generally provide trunk and regional service, with the short takeoff and landing aircraft (STOL) and vertical takeoff and landing aircraft (VTOL) providing shorter haul service between and within urban areas. Air traffic control routings and instructions assure three-dimensional separation between the flight paths of the various airspace users: air carrier, general aviation, and military.

ascent and descent. The airborne equipment used by the v/STOL's (even though having some special features), must be compatible with whatever collision avoidance or separation system may be used by CTOL aircraft.

v/STOL aircraft, because of their capability to change speed readily over wide ranges, lend themselves particularly to a traffic control technique incorporating some form of planned air-to-air, self-separation (station-keeping) between aircraft in the same flight environment by means of a direct airborne readout to the pilot, but used as part of the overall ATC System. Such a concept would increase airspace utilization by permitting close spacing vertically, laterally, and longitudinally en route. It would also provide for close separation between v/STOL's while landing and taking off.

Communications

In the conventional Air Traffic Control System, a great volume of communications between pilots and controllers is involved, not only during en route flights, but particularly during the approach pattern. These are both for navigation and to prevent collisions between aircraft, based primarily on interpretation by the controller of flight data derived from ground radar equipment. Voice communications required by ATC to handle high volume v/STOL operations in the conventional manner not only would be physically impossible (due to frequent landings, takeoffs and altitude changes, plus complicated flight tracks), but would require excessive numbers of ground-control personnel and radio frequencies, as well as being a distracting burden to the pilot during critical flight maneuvers. Thus, automatic—data link—communications become a *must* for high-density v/STOL traffic.

All-Weather Operations

In order that v/STOL aircraft can fulfill their logical role in providing reliable short-haul transportation, they must be capable of operating under all-weather conditions, day and night. Weather minima are classified in several categories for operational purposes as discussed in more detail in chapter 10. These are applicable to all types of aircraft and are summarized as follows.

	Landing decision height in feet	Runway visual range (RVR) in feet
Category I	Down to 200	Down to 2,600
Category II	200 - 100	2,600 - 1,200
Category III A	Under 100	700
Category III B	Under 100	150
Category III C	Full instrument landing and taxiing (zero-zero)	

The majority of air carrier and other CTOL IFR operations are in Category I, but increasingly they are moving into Category II. By the late 1970s and early 1980s, most operations at major hub airports are expected to be in Category III. v/STOL airline-type operations, accordingly, will at least have to keep pace in order to provide short-haul transportation on the same basis of reliability in order to serve as part of a coordinated air-transport system.

v/STOL aircraft have the potential for operating to the same or lower weather minima as CTOL aircraft, because of their ability to fly at slower speeds. In addition, they have characteristics which permit better utilization of airspace and thus provide capability to achieve a greater volume of air traffic in the same relative environment, as discussed earlier. To fulfill these potentials, however, the ATC System must be designed and operated in a manner such that v/STOL's are able to utilize to the fullest extent their low-speed, flexible flight characteristics and all-weather capability.

PERSONAL AIRCRAFT

Over the years, many efforts have been made to develop an air vehicle that also could be operated on streets and highways and thus meet personal transportation requirements in a manner superior to the automobile. Technology and economics, however, so far have not been able to produce a viable product. The automobile has continued to be dominant to an overwhelming degree in the field of personal transportation.

But today, the automobile—and the surface—are in serious trouble! Traffic jams in peak hours slow the automobile down to about as fast as a person can walk. Highways, roadways, and streets cover unacceptably large pro-

portions of the surface in relation to that available for living in urban areas. Collisions between automobiles produce an alarmingly high accident death rate. The virtually untapped capability of the *airspace* must be considered seriously as a medium to alleviate the surface congestion problems for personal—private—mass transportation.

Probably the most significant single factor in determining whether the personal or general aviation growth rate could more closely parallel that of the automobile is the question of the utility of the airborne vehicle. In the case of the conventional fixed-wing aircraft, the utility has been quite limited. The owner is still required to drive to the airport; the airplane has to be maintained at the airport away from his residence. At the end of a trip, some sort of surface transportation is required to carry on after landing at the airport of destination. In effect, the conventional aircraft does not have enough utility to satisfy the average family. Should an airborne vehicle serve to provide personal transportation on a more or less "home-to-office-to-home" basis, however, the picture certainly would be changed radically.

One of the problems to solve in this respect relates to the psychological makeup of human beings, who generally may react against the idea of approaching the ground at relatively high speeds. While approaching the ground, they intuitively realize that what they do or fail to do at high speed is at that moment affecting the ultimate outcome of their flight, their own safety, and that of others. Reflexes of the average human are such that he just does not always react simultaneously to an emergency or to a critical situation requiring an immediate decision. A simple-to-operate vehicle which would permit the operator to slow down, and if necessary pause to think, might very well overcome these psychological and reflex problems. If, in fact, the vehicle could approach the landing spot at a very slow speed, or vertically, considerable appeal to the average person with an automobile-driving conditioning might be expected.

The question of dealing with high-density air traffic, such as might be produced by the introduction of a utility/personally operated air vehicle, is only *relative* to the surface-traffic problem. It is obvious that surface traffic with automobiles cannot increase indefinitely.

Although there are problems in increasing air traffic density, the three-dimensional capability of airborne vehicles, as against the two-dimensional capability of surface transport, may offer a very significant solution to future transportation problems in general.

Highways, roads, and streets have a finite capacity; *airspace*—**The Uncrowded Sky**—has an infinite capacity.

THE SUPERSONIC TRANSPORT AIRCRAFT

During the mid to late 1970s, two types of supersonic transport (SST) aircraft are expected to be in operation. One is the *Concorde*, manufactured by France and England on a joint-venture basis; the other, the TU-144, by Russia. Development of a United States SST—canceled in 1970—will probably be resumed during the late 1970s. Some experts forecast a probable free-world SST market of 400 aircraft by the 1980s and some 800 to 1,000 by the 1990s.

Because of its operation in both higher speed and greater altitude regimes, the SST becomes an important factor in the ATC System. The system must recognize the different problems which this type of aircraft will generate and provide solutions that will not impair its special operational characteristics such as critical fuel considerations, particularly during the climb phase.

The SST is an aircraft that can fly faster than the speed of sound, identified as Mach 1, which is approximately 660 mph or 573 knots at 36,000 feet. The term "Mach" is derived from the name of Dr. Ernst Mach, an Austrian scientist who pioneered research into the problems of supersonic flight in the late nineteenth century. The first generation SSTs will fly between Mach 2.2—1,450 mph or 1,260 knots—and Mach 3—1,980 mph or 1,720 knots. Perhaps the Mach 3 designs will be capable of a later increase in speeds to approach Mach 4—2,540 mph or 2,300 knots.

Operational Altitudes

The SST's operate at maximum speeds in the stratosphere mainly between 50,000 and 75,000 feet; however, they

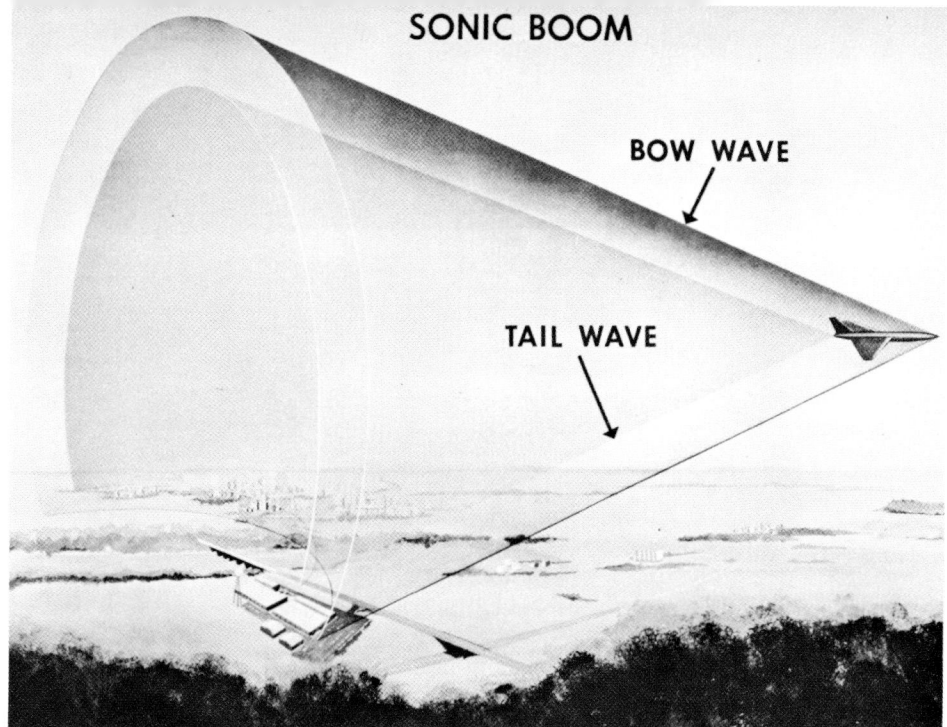

126 Sonic boom trails SST with shock waves sweeping across the surface.

127 French/English SST.

must also be able to operate economically at lower altitudes and fit into air traffic patterns at speeds compatible with those of the subsonic aircraft, when required. Operation at supersonic speeds will not be performed under 40,000 feet.

Range

In general, the SST will not be able to operate economically in competition with subsonic jets under stages of about 2,000 nautical miles. The optimum range will be between 2,000 and 4,500 nautical miles. Depending on stage lengths, around 60 to 80 percent of the distance will be flown at supersonic speeds; the remainder will be at subsonic speeds during climb, descent, and approaches for a landing.

Airports

By international agreement, all SST's either designed or to be designed must be capable of using the same runways and taxiways as the conventional, large, subsonic jet airliners.

Sonic Boom

The sonic boom has been considered one of the most serious problems associated with the introduction of SST's in civil air transportation. In order to attain the goal of publicly acceptable operation of SST's over populated areas, the sonic boom is a fundamental consideration in the mission specification, airframe/engine design, and flight operation. Practical methods for reducing the shock-wave intensity to tolerable values other than through restraints on the aircraft design and operation must be found.

A sonic boom is produced by the rapid pressure changes caused as shock waves, radiating from the aircraft in supersonic flight, sweep across the ground. The intensity of the pressure change (or "over-pressure") is dependent mainly on the weight (and shaping) of the aircraft and the altitude at which it is flying. The heavier the aircraft, the greater the over-pressure it is capable of producing; and the greater the altitude, the more the over-pressure at ground level will be reduced. Although of short duration at any given point, the sonic boom "carpet" commences with acceleration through about Mach 1.13 and ends with deceleration through the same speed.

Certain supersonic maneuvers, such as turns, pushovers in an altitude change, and acceleration at a high rate, can result in pressure buildups on the ground that are commonly referred to as "superbooms." For example, a turn at supersonic speed can amplify the boom effect two to four times. These superbooms are momentarily fixed and do not move with the aircraft; they occur over a relatively small expanse on the surface, approximately one square mile.

Airport Noise

SST aircraft will follow the same "noise abatement" flight procedures at airports as are established for the CTOL subsonic jets. Although somewhat noiser on takeoff than subsonic jets, due to the higher power of its engines, the SST will probably be quieter in climb since its large reserves of thrust will permit partial throttling very soon after takeoff. At 3.5 miles from start of takeoff—a commonly used noise level reference point—the SST can be quieter than the large subsonic jets, because its steeper climb angle will enable it to reach a higher altitude at that point.

All-Weather Operation

At supersonic speeds, the SST is particularly susceptible to meteorological phenomena such as turbulence, hail, ice, and perhaps even rain. The most important factor is temperature. Special meteorological forecasting services are required for SST operation; however, the probability of turbulence at supersonic altitudes is about one-third that at altitudes where most of the subsonic jets normally fly; gust intensity is about one-half. Initial SST operations can be carried out down to Category II minima. Eventually, the aircraft and the airports they use will permit operations in Category III conditions.

Flight Profile

A first step in appraising the SST effects on the ATC System is to consider a hypothetical flight profile (based on international standard atmospheric conditions) of a Mach 2.2 aircraft. This will indicate the type of flight situation

128   Russian SST.

which will have to be handled by ATC in the early SST era.

From takeoff (assuming sea level) it will take 5 minutes to reach 20,000 feet at about 35 nautical miles from the takeoff point, and another 3 minutes to reach 36,000 feet at about 70 nautical miles. During this period the average true air speed will be about 400 knots and the rate of climb 5,000 to 8,000 feet per minute. From this point the aircraft enters the transition phase from subsonic to supersonic speed and cruising level, and the acceleration from Mach 0.91 to Mach 1.5 will take about 4 minutes and about another 60 nautical miles. Speeds during this phase are around 590 knots at 30,000 feet, 820 knots at 40,000 feet, and 1,130 knots (Mach 2) at 50,000 feet. Total time from takeoff to Mach 2.2 at a cruising altitude of at least 54,000 feet will have taken 16 to 20 minutes over a distance of 220 to 300 nautical miles.

Fuel Considerations

During the climb of an SST to supersonic cruising level, about 20 percent of its fuel will be consumed under ideal conditions. If, on the other hand, a normal ascent flight profile is not followed because of significant ATC or other interference, the fuel consumption could go so high that flight could not be continued and the aircraft would have to return to point of departure before ever reaching supersonic speed. A 15 to 20 minute delay in reaching supersonic speed could double normal fuel consumption during ascent and could well mean an aborted flight. For example, undue restrictions on departure for "sonic boom control" or "noise abatement" could seriously interfere with the ideal ascent profile of the SST. Other interference could be caused by inability of ATC to fit an outgoing SST into a congested traffic area, thus requiring maintenance of certain subsonic speed altitudes for significant periods of time, or requiring that the SST follow extensive and complicated departure paths at subsonic speeds.

Similar fuel consumption considerations exist in connection with the descent flight profile, and any undue holding of the SST in traffic congested situations would be even more serious as the aircraft in excessive delay conditions could rapidly use its reserve fuel supply.

SST Maneuverability

At cruise speeds, the turn radius of an SST is around six times that of a subsonic jet. To illustrate, at Mach 2.2, assuming constant speed and a 30° bank, the turning radius is about 40 nautical miles; at a 20° bank, the turning radius is about 65 nautical miles. A heading change of 45° will take three to four times longer for an SST than for a subsonic jet.

At transonic speeds (between Mach 0.91 and Mach 1.5) the SST has greatly reduced operational capability. Turns and leveling in this speed range, i.e., between 40,000 and 50,000 feet altitude, would not be contemplated due to the SST's minimum performance during this phase of its flight profile; for example, a 25° bank during a climb at transonic speed may reduce the rate of climb to zero. During a climb of over 50,000 feet, the lead time to level is about 10,000 feet; in descent from above 50,000 feet, the lead time to level is about 5,000 feet.

While at subsonic speeds, the SST has about the same maneuverability characteristics as the large subsonic jets; however, lead time to level in a climb is about 5,000 feet, and in a descent it is about 3,000 feet.

Special ATC Factors

The operational characteristics of the SST present a number of special challenges to the Air Traffic Control System.

One factor which the ATC System must take particular notice of is that of fuel considerations. Since the major design criterion of this aircraft concerns speed, the engines have to be large, which, in turn, makes the economical operation of the aircraft extremely important. Because of this factor, delays on the ground as well as in the air must be kept to a minimum.

Another factor which directly affects the control of air traffic is the high speed of the SST. The ATC System must be prepared to handle communications very rapidly, particularly when transferring control responsibility, since the aircraft will pass through control areas very quickly.

Since the SST will be operated at subsonic speeds in terminal areas, the ATC System must be prepared to integrate its operations with those of other aircraft.

CHAPTER X

129 Model of proposed United States SST.

AIRPORTS

Some greater than conventional airspace reservation will be needed for the SST in subsonic speed due to its ascent and descent characteristics and the longer lead time needed for leveling than is required by subsonic jets.

Another ATC consideration is that routes and air traffic control procedures must be designed in a manner such that SST's will be able to avoid, to the maximum extent possible, turns, rapid acceleration, and abrupt altitude changes both at supersonic and at transonic speeds.

The high speeds of the SST's emphasize the need for substantial automation in Air Traffic Control—automated communications between control facilities, automated flight-data processing, more efficient displays for controllers, automatic position reporting, and a substantial degree of ground-air-ground automatic communications. The importance of some form of air-to-air separation techniques by means of air-derived separation-assurance airborne equipment also is emphasized. This is not only because of the SST's high speeds, but also because the pilots will have very restricted visibility except in the relatively short periods before landing and after takeoff when the SST's hinged nose is in the lowered position.

THE HYPERSONIC TRANSPORT AIRCRAFT

Looking ahead to a possible successor to the SST, it may be that the hypersonic transport (HST) provides an obvious follow-on aircraft. One approach to such an aircraft is based upon the use of ramjet engines to attain a speed of Mach 4.5-5. To reach Mach 6, a "supersonic combustion ramjet" (scramjet) would be required. While these engines would cost up to twice those for an SST, their fuel consumption would be considerably lower, due to their greater efficiency. Unless speeds above Mach 6 are required, the airframe technology used in the SST should be adequate, although the cost would be about twice as high. An HST could be made available in the 1985-1990 period.

Both the SST and the HST are practicable only for long distance flights—the SST becomes useful at any distance over 1,200 miles. The HST only shows to advantage at distances of around 4,000 miles, which would appear to mean that they could be operated simultaneously.

Since all air traffic originates and terminates at some point on the earth's surface, the overall efficiency of the Air Traffic Control System is directly affected by the adequacy of the landing/takeoff areas, or "airports." Improvements in airports thus must be undertaken constantly as part of overall improvements in the ATC System, and in keeping with the development of new aircraft and expanding air traffic volume. Unless airport progress keeps pace with all of the technological advances in aviation, "airports" can become one of the most serious bottlenecks in the path of efficient and safe air transportation.

As a matter of fact, airport capacity generally does not keep up with the demands of the world's increasing air traffic. This deficiency has the direct effect of causing delays in the arrival and departure of aircraft which can vary in proportion to an airport's particular deficiencies. An easy solution to the delay problem is to limit the volume of air traffic which can use the airport under specified conditions or periods of time; but this is a "bandaid," not a cure for the problem.

Delay of air traffic movements is not the only factor, however, in measuring the adequacy of airports. Another consideration is the efficiency of an airport and related facilities in handling the flow of passengers and cargo to and from the arriving and departing aircraft.

The ability of an airport to provide for reliable air service in adverse weather conditions also is important.

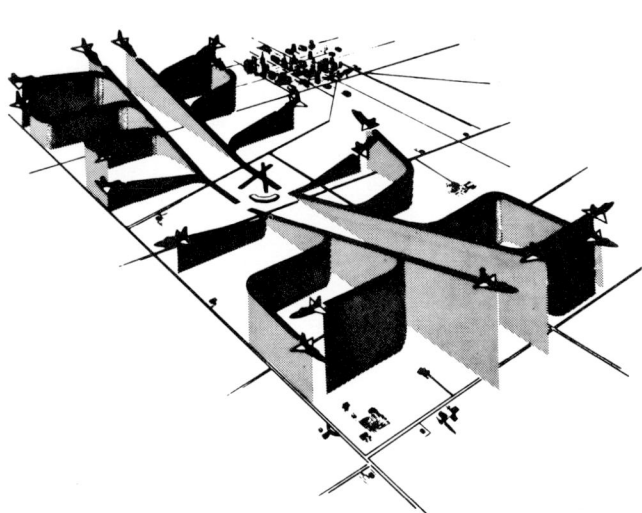

130 *The challenge: maximizing airport capacity.*

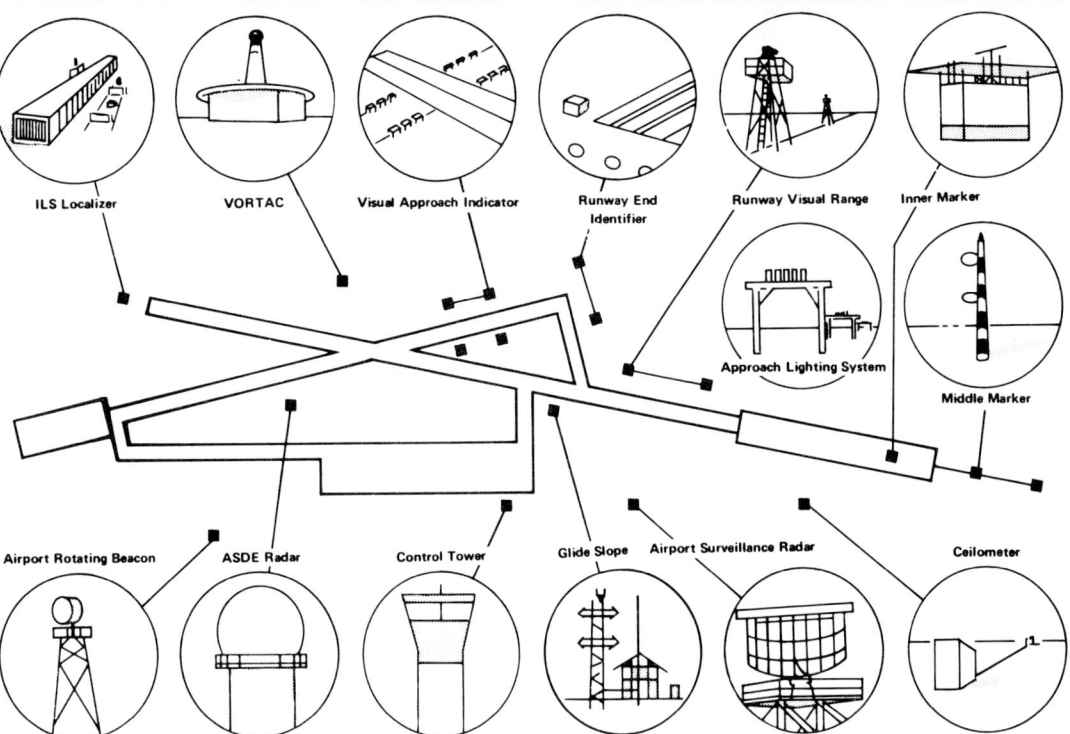

131 *Typical airport layout and location plan for related ATC and navigation facilities.*

This capability is needed not only for large metropolitan terminals, but also for airports serving small population areas.

Airports—of whatever type or class—as an integral part of the ATC System must provide adequacy of service such that air transportation can bring its vital contributions to economic development and progress to the greatest number of people throughout the world.

AIRPORT FUNCTION

The airport has become a focal point in urban economic growth, and its importance as an urban growth factor will continue to increase. Historically, cities and towns have developed around the transportation terminals which were in the greatest use at a given time. This concentration has varied from seaports and riverports to railroad terminals to air terminals. In many locations, all may be growing along with the population.

Some large metropolitan airports have highly developed commercial and industrial complexes nearby. Airport industrial parks are constructed especially at general aviation airports where corporate aircraft have access directly to their business establishments. There is an increasing demand for air-cargo facilities at airports, such as terminal storage areas, air-express facilities, and the capability to handle motor-carrier traffic which has been generated by the air-cargo facility.

In order to provide for such a complex, the capability of the airport to facilitate the operation of large numbers of aircraft of all categories, under all types of weather conditions, is extremely vital. Provision must be made for approach and landing aids, both electronic and visual. Local and area air traffic control service is, of course another must. Runways and taxiways, as well as other landing/takeoff areas, must be adequate to handle the classes of aircraft which will use the airport. Supporting equipment, such as snow removal and fire fighting, needs to be provided. Parking areas for aircraft and automobiles must be sufficient to handle the anticipated traffic. These important factors apply to any airport, regardless of its size or use.

Due to the large number of passengers—as well as those who meet them—which use a modern metropolitan airport, a great deal of care must go into the planning for facilities adequate to handle the volume efficiently. A number of different approaches have been used in attempts to cope with this situation. The most usual method is a single "terminal" building, through which all passengers and their baggage are routed. Ticket offices, stores, and other amenities are also located in these buildings. Another approach has been to provide a number of "satellite" buildings to carry out the same function. In this case, each satellite terminal must be provided with all of the passenger facilities. A completely different approach uses buses or "lounges" to take the passenger to the aircraft from a single main building where ticketing and baggage are handled.

In general, whatever the design of the passenger handling facilities, most of these "terminal" buildings are a bottleneck, at least during peak traffic periods. As was discussed previously, a system whereby passengers are flown to the airport directly from a city center by a V/STOL aircraft, perhaps carrying a bus or "pod," appears to offer promising possibilities to decrease terminal-building congestion and make possible a much more efficient flow of passengers, cargo, and baggage through an airport.

AIRPORT ELEMENTS

The elements which comprise the overall airport complex are:
- Access/egress facilities.
- Airport landside which consists of passenger and cargo-handling buildings, including aircraft loading and unloading facilities; also the taxiways and runways with their related facilities such as lights and landing aids.
- Airport airside. This element relates to the capacity and capability of the airport to accommodate different categories of aircraft: CTOL, RTOL, STOL, VTOL.
- Terminal airspace. Involved in this element is the airport's accessibility, airspace restrictions to and from the airport, traffic-pattern conflict with other neighboring airports, and airspace structuring to facilitate arriving/departing aircraft.

132 A major airport needs efficient supporting surface transportation.

The last three elements enumerated above are of direct concern in the control of air traffic. The characteristics of each of these three elements not only determine an individual airport's capabilities, but in the aggregate for all airports they also affect the capacity and efficiency of the total ATC System.

AIRPORT PLANNING

The worldwide system of airports has developed as a reaction to the growth in the use of air transportation. This continual expansion in air transportation has been followed by an increase in the number of conventional airports for CTOL aircraft as well as increases in the size and character of the related "jetports."

As CTOL air-traffic density increases and produces operational delays and airport congestion problems, an obvious solution may appear to lie in the provision of additional conventional airports, strategically located so as to absorb some of this traffic. But real estate for conventional, large jetports has become less and less available and more and more expensive, particularly in the densely populated areas where additional air service is needed. New major jetports will, of course, be built in some selected parts of the world. "Regional" airports removed from but serving one or more metropolitan areas are included in this category.

On the other hand, the capacity of existing CTOL airports can be increased by several methods. These include

133 One of seven terminal complexes for Dallas-Fort Worth regional airport.

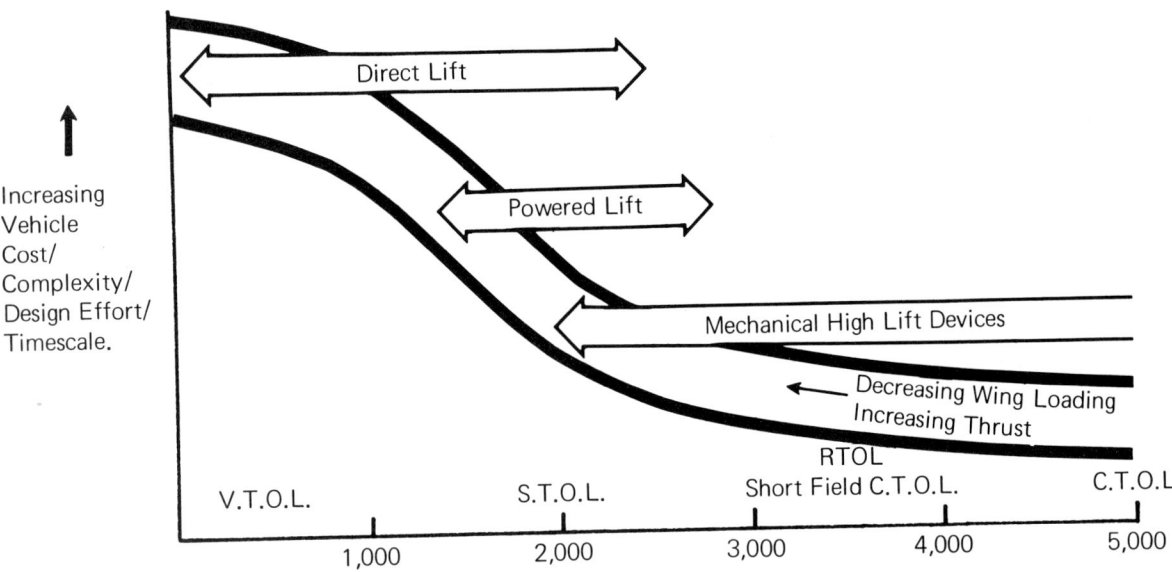

134 Effect of aircraft design on short-field length (in feet).

improving runway/taxiway design and layout, providing parallel runways, and increasing runway utilization rate through improved ATC spacing techniques.

Segregating different classes of aircraft is another factor involved in airport planning. Certain runways may be reserved for high performance aircraft used by general aviation and for short-haul CTOL air services. Further, facilities to handle STOL and VTOL aircraft may be constructed on jetports to "unload" CTOL traffic.

Additional unloading of congestion at jetports may be accomplished through the distribution of various categories of air traffic to specialized airports. These can include general aviation airports, executive/business airports, cargo airports, STOLports/RTOLports and VTOLports.

Effective airport planning along the foregoing lines can provide virtually unlimited capacity to the ground facilities from which "all air traffic originates and terminates." The design and operation of the ATC System will need to keep pace with the inevitable growth in airport productivity which will take place as the demand for air transportation services continues to increase.

OPERATIONAL WEATHER CATEGORIES

Standards have been established by the International Civil Aviation Organization as a guide to member countries for classifying landing operations applicable to all categories of aircraft in accordance with specified weather minima.

Category I: Operation down to minima of 60 meters (200 feet) decision height and 800 meters (2,600 feet) runway visual range (RVR).
[*Note:* "decision height" refers to altitude above the ground at which the pilot will decide whether the approach can be continued with a high probability of successful completion; if not, whether another approach should be attempted or the flight shoud proceed to an alternate airport. "Runway visual range" (RVR) is an instrumentally derived value that represents the horizontal distance a pilot will see down the runway from the approach end. The primary instrument used to determine RVR is the transmissometer. It consists of a projector, a detector, and a meter to indicate the transmission of light through the atmosphere. The meter converts this transmissivity into a measure of visibility which is extrapolated into RVR values up to 6,000 feet. Readouts can be taken from the transmissometer as desired at various locations at any airport.]

Category II: Operation down to minima below 60 meters (200 feet) decision height and 800 meters (2,600 feet) RVR, and as low as 30 meters (100 feet) decision height and 400 meters (1,200 feet) RVR.

Category III A: Operation to and along the surface of a runway with external visual reference during the final phase of the landing to an RVR minimum of 200 meters (700 feet).

Category III B: Operation to and along the surface of the runway and taxiways with visibility sufficient only for visual taxiing comparable to an RVR value in the order of 50 meters (150 feet).

Category III C: Operation to and along the surface of the runway and taxiways without external visual reference.
Pilot requirements: The operational weather categories described above may be utilized only after the pilot possesses the appropriate authorizations and ratings to meet the criteria specified for each such category, by the proper authority of the country which licensed the pilot.
Aircraft requirements: The aircraft equipment required to operate in each of the above operational weather categories is specified by the appropriate authority of the country in which the aircraft is registered.

Airport requirements: In order for an airport to be used under the above operational weather categories, it must be so designated and be provided with such equipment and facilities as may be specified by the appropriate authority of the country in which the airport is located.

CTOL AIRPORTS
Runway Capacity

The basic standard for determining runway capacity or utilization applicable to CTOL aircraft is the principle that an arriving aircraft should not be committed for a landing and that a departing aircraft will not commence take-off until a preceding aircraft has turned off the runway or is completely airborne. What this means is that there must be a minimum spacing between aircraft which physically limits the number of aircraft that can be accommodated by a given runway in any given time period.

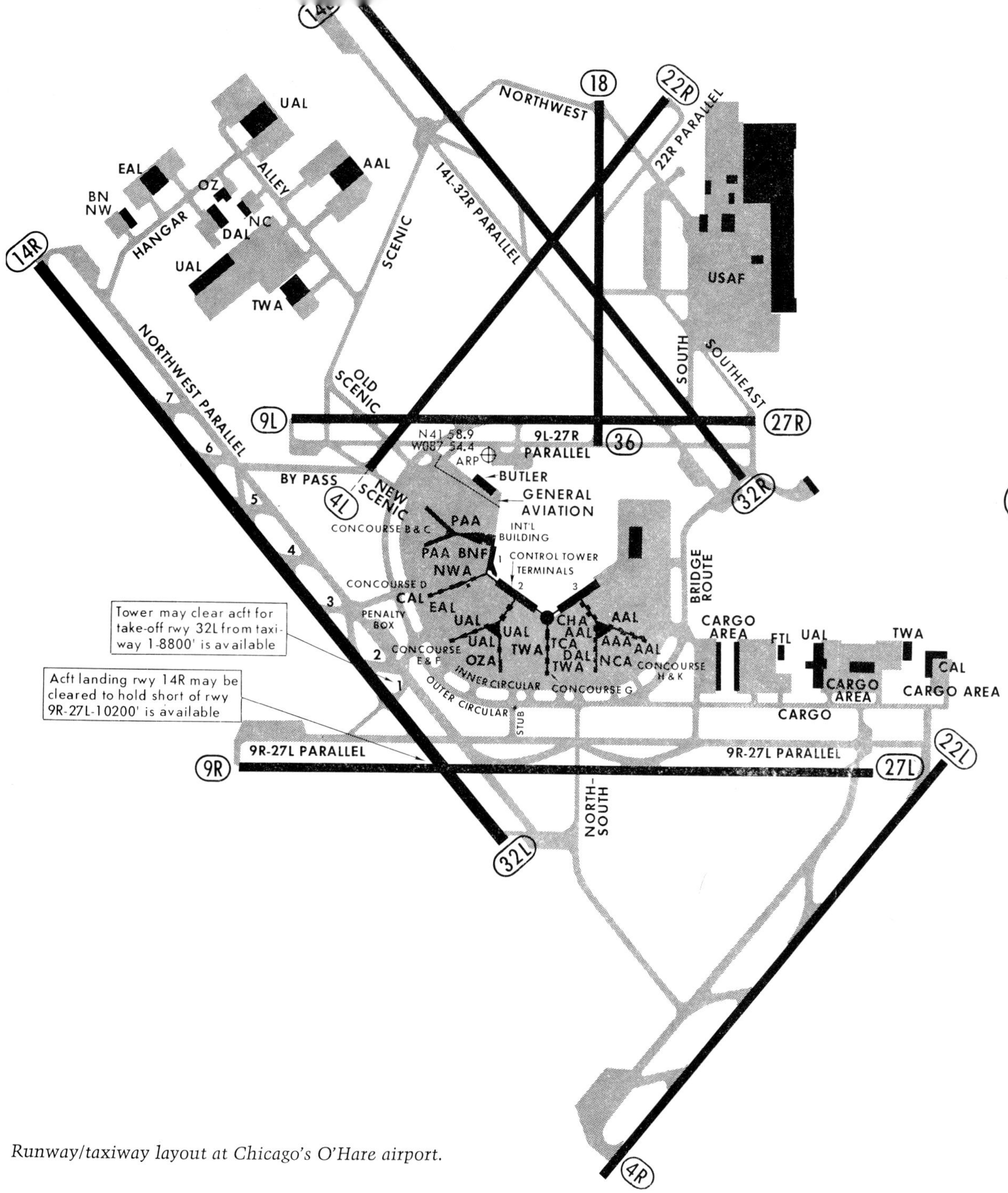

135 Runway/taxiway layout at Chicago's O'Hare airport.

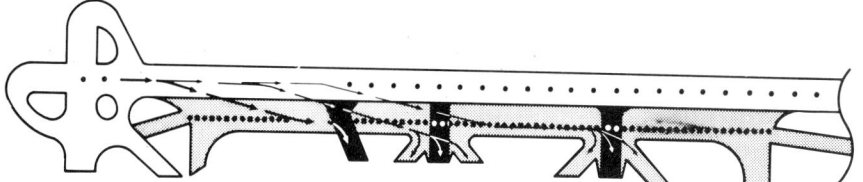

136 Drift-off (deceleration) area includes existing taxiways and permits the pilot to vacate an active runway rapidly, thus increasing runway capacity.

In actual practice, maximum runway utilization at the busiest conventional jetports averages about 40 to 45 movements (landings and takeoffs) per hour under optimum conditions. The *theoretical* capacity of a runway when used by high-performance CTOL aircraft in mixed takeoff and landing conditions is at least 80 movements per hour, based on *physically* feasible 45-second intervals). In other words, CTOL runways are being used to not more than 50 percent of their potential capacity even at the high-density jetports. If this is true, what can be done?

Achieving Maximum CTOL Runway Capacity

Obviously, it is desirable for economic reasons to achieve a runway's maximum capacity to the greatest extent possible before investing the large sums involved in providing additional CTOL runways at an existing airport or in constructing additional airports. Certain steps can be taken in this direction under two broad categories: design factors and ATC factors.

Design factors: The first and perhaps most significant single step in this category is the provision of high speed turn-off or exit taxiway systems. These should be bi-directional, i.e., usable for aircraft landing in either direction on a runway. They should be designed at a small angle—rather than at 90°—so that at the higher landing speeds, aircraft can turn off conveniently at 1,000 foot intervals, beginning about 2,000 feet from the approach end of the runway. Carrying out this design factor provides the means for a landing aircraft to make a quick exit from the runway and thus free it for the next aircraft.

Perhaps a more effective method of decreasing runway occupancy time is to connect parallel taxiways to the runway with a "deceleration" or "drift-off" area. This would permit pilots to leave the runway immediately after landing, at the point where the aircraft has reached a speed at which it may safely make a turn. Such an arrangement gives to the pilot and/or the controller the flexibility to clear the runway in use at the earliest possible moment, thus permitting another aircraft to land or takeoff immediately. This enlarged combined runway/taxiway would be suitably marked and lighted to delineate the runway edge, as well as the taxiway. This improvement alone could be expected to radically increase the capacity of a runway. It would eliminate the delays normally experienced through the necessity for a departing aircraft, for example, having to delay its takeoff until the landing aircraft has found the assigned turn-off taxiway, slowed down, and safely cleared the runway.

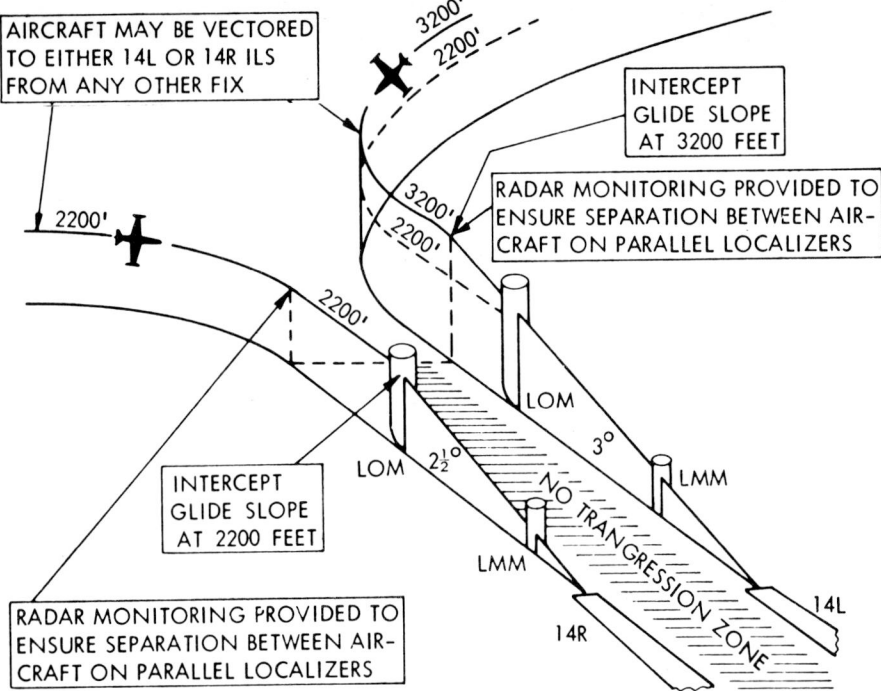

137 Simultaneous landings on parallel ILS runways contribute materially to increased airport capacity during IFR conditions.

138 Control tower at Toronto International Airport.

A third important design factor for improving runway efficiency is the provision of run-up pads (holding aprons) large enough so that a departing aircraft may pass another which has been delayed prior to takeoff, thus eliminating takeoff "jams" and permitting uninterrupted departing traffic flow during peak periods.

While the above steps may appear simple, very few airports in the world incorporate all of these features. Probably a 20 to 30 percent increase in capacity in any given runway configuration could be achieved by complying with one or more of these three runway design factors.

ATC *factors*: The goal of the Air Traffic Control System is to permit aircraft to land and takeoff — in all-weather conditions — at the maximum physical capacity of each airport (or other landing/takeoff area). However, virtually every conventional airport has a higher total handling capacity during VFR conditions than during IFR. This is because, in general, the airport's *total* runway configuration is used in VFR, but only selected "instrument" runways during IFR. The ideal ATC system should be capable of bringing an airport's IFR capacity up to that of its VFR capacity.

Achieving maximum *runway* capacity revolves around the ATC system's capability for providing consistently close, yet safe, air-to-air separation between aircraft — both while approaching for a landing and after takeoff. For example, if a runway's design is such that landing aircraft can vacate the runway in 45 seconds after arrival at the runway threshold, then arriving aircraft could be spaced at 2 nautical-mile intervals (assuming an approach speed of 160 knots). If 60 seconds are required to vacate the runway, then arriving aircraft (at the same speed) would be spaced at about 3 nautical miles, which is the generally accepted *minimum* longitudinal spacing between landing CTOL aircraft in the third (mid-1970s) generation ATC system.

Consistently precise spacing of arriving aircraft, however, cannot be accomplished by the controller with visual estimates, either by use of radar or by seeing the aircraft from the control towers. Neither can the pilot maintain precise spacing in the approach and landing patterns by visual estimates, even in the best of weather. Both need computer assistance — ground and airborne — to achieve these objectives. This involves the use of "time-referenced" navigation, as discussed earlier.

To facilitate optimum spacing during approaches and landings, aircraft should be able to maintain relatively similar speeds. This is no great problem for many of the larger piston-powered aircraft, the subsonic jets — both airline and general aviation — or even the SST's. Smaller aircraft which cannot maintain an approach speed within acceptable tolerances, however, cannot fit into maximum density patterns using the same runway with larger aircraft, if maximum runway utilization is to be achieved.

For departing air traffic, speed characteristics of different aircraft are not as critical, provided that ATC can give separation through lateral, (parallel track) spacing techniques. Spacing, preferably lateral, between departing and arriving aircraft must be so arranged that all traffic can move freely to and from the runways at their maximum flow rate. Also, ATC must be able to fit departing aircraft into the en route traffic patterns whenever departing aircraft are ready; otherwise, departures are delayed and runway capacity is reduced from its theoretical maximum merely because of ATC problems in the upper airspace. The use of area navigation to permit the pilot to follow pre-organized flight profiles in three dimensions (3-D RNAV) facilitates the solution of spacing problems for both the controller and the pilot.

Longer range system improvements which can help bring the capacity of a properly designed CTOL runway to its maximum theoretical capacity, 80 movements per hour (or more), in all-weather conditions, include:

- Improved flow control and optimum speed determination derived from the ATC computer complexes.
- Provision of airborne air-to-air spacing equipment with direct pilot readout.
- Acquisition of flight data by an automatic digital (data link) communication system.
- Data link controller/pilot communications.

In approaching the maximum physical capacity of a runway, of necessity, there may be a small delay for each aircraft in order to achieve a "cushion" to permit a con-

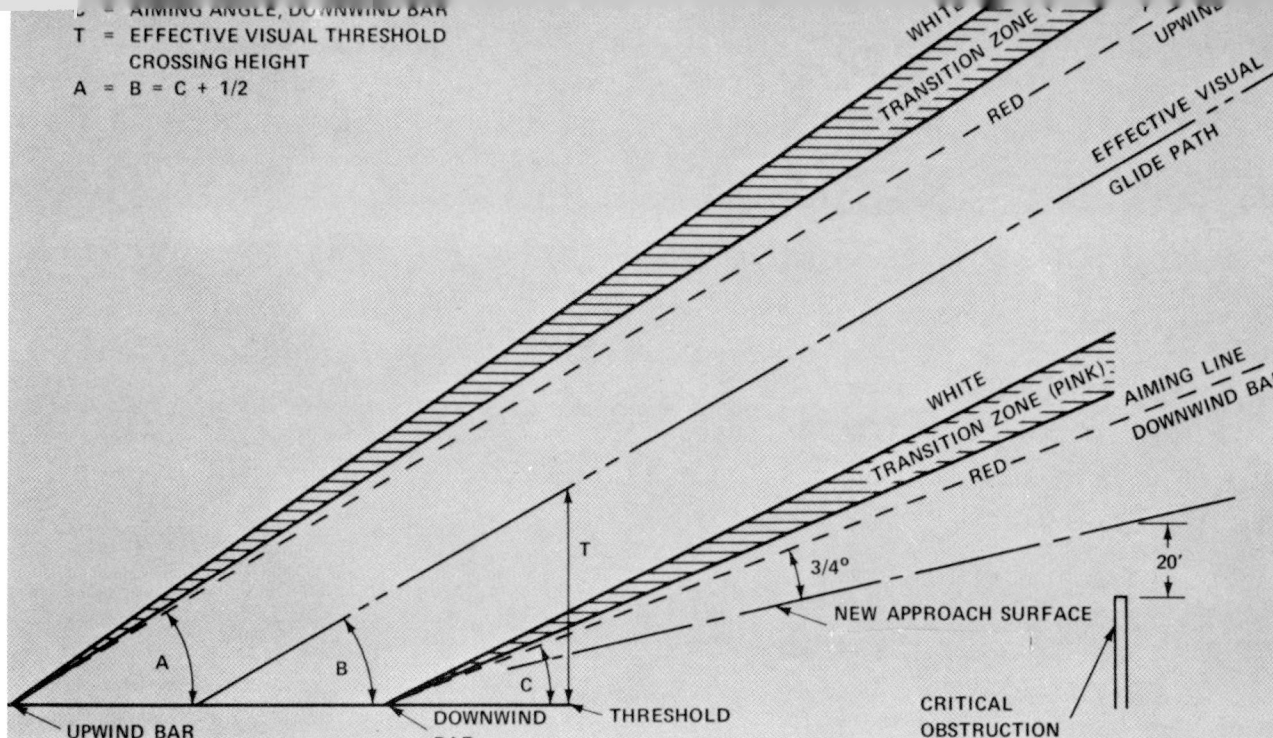

139 Approach lighting system (VASI) provides pilot with visual glide slope during final segment before landing.

stant flow of traffic. This would be something in the order of 3 to 5 minutes on the average, with arriving aircraft absorbing the delay by reduced speed and/or "stretched" approach patterns; departing aircraft would take the delay while taxiing to the takeoff point.

LIGHTING

An essential part of any major conventional airport to facilitate air traffic use day and night and under all-weather conditions, is its lighting system. International standards for such lights have been established by ICAO which call for certain minimum equipment. Typical of standard installations throughout the world are those provided at major airports in the United States. Some of those normally installed are:

- Rotating beacons to mark the location of an airport.
- Instrument-approach light system to provide the basic means for transition from instrument flight to visual flight and landing.
- Condenser-discharge sequenced flashing-light system which is a series of brilliant blue-white bursts of light flashing in sequence along the approach lights.
- Visual-approach slope indicator (VASI) which provides visually the same information that a glide-slope unit of an ILS provides electronically, in the form of red and white lights to indicate the correct glide path.
- Touchdown zone and runway-centerline lighting to facilitate landing under adverse visibility conditions.

- Runway-end identifier lights (REIL) to facilitate rapid and positive identification of the approach end of a runway.

NOISE

Another factor which affects the operation of an airport from an air traffic control standpoint is aircraft noise and its relationship to the airport community. Highly populated residential areas and housing units, such as large apartment buildings, exist in the vicinity of most major airports. This, in many instances, has brought about serious conflicts between aircraft/airport operators and the residents. Some airports are closed during certain hours of the night only because of the noise problem. "Noise abatement" procedures are applicable at every major airport, and these can seriously interfere with efficient air traffic flow.

Noise abatement procedures apply to both departing and arriving CTOL jet aircraft in particular and, to a certain extent, to all aircraft. Noise abatement for both arriving and departing aircraft have one procedure in common. When at lower altitudes, generally below 3,000 feet above the surface, they follow approach and departure flight paths which avoid highly populated areas to the greatest extent possible. This may involve one or more turns, for example, right after takeoff or just before landing. Also, when conditions permit, ATC may use a runway or runways which direct air traffic away from residential districts.

Two somewhat diametrically opposed noise-abatement procedures may be followed by departing CTOL jet aircraft. In the first, the pilot immediately after takeoff reduces power to the minimum safe speed until past residential areas before resuming normal climb power. In the second, which generally is followed in the United States, the pilot maintains maximum climb power and maximum climb gradient from takeoff to 3,000 feet above the surface, and thereafter proceeds at normal climb power and gradient. Aircraft above 3,000 feet are not considered a major noise source.

In the case of arriving aircraft, a noise-abatement procedure involves the use of a lower landing flap setting whenever possible and a lesser approach flap setting throughout the approach. By using lesser flap settings, drag is reduced and a lower power setting is required to maintain a steady descent.

A different noise abatement procedure for arriving aircraft involves the use of a "two-segment approach." In this type of procedure, the pilot follows a relatively steep first-segment descent gradient (5-6°) to a predetermined distance from the end of the runway and to a predetermined altitude. At this point, transition is made to a normal 3° second-segment descent gradient to landing, thus keeping the aircraft at higher altitudes until close-in to the airport. As a result, low altitude noise is reduced to a minimum over residential areas. The pilot may perform a two-segment approach by using an airborne

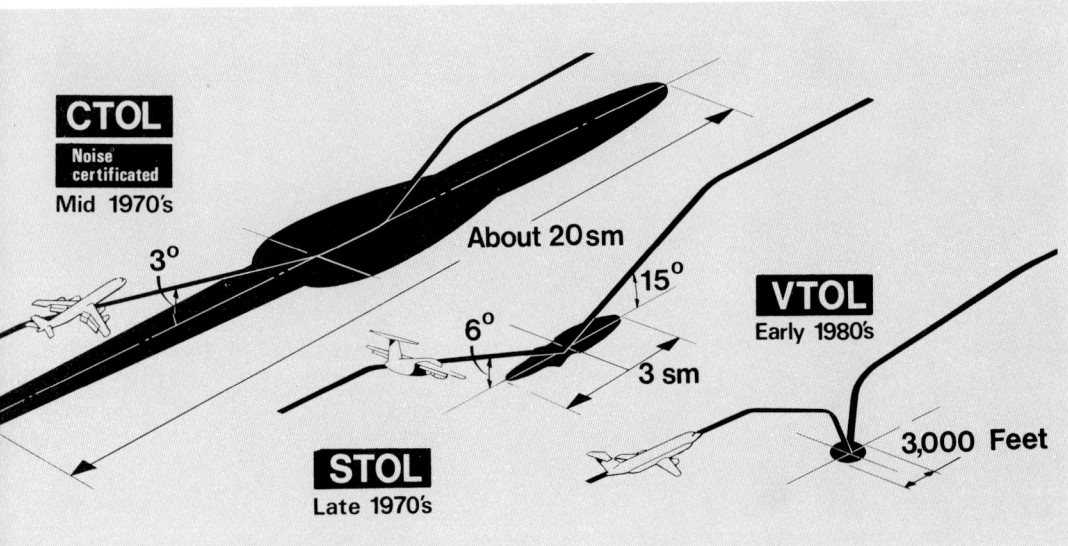

140 Comparative 90 EPNdB noise footprints for CTOL, STOL, and VTOL aircraft.

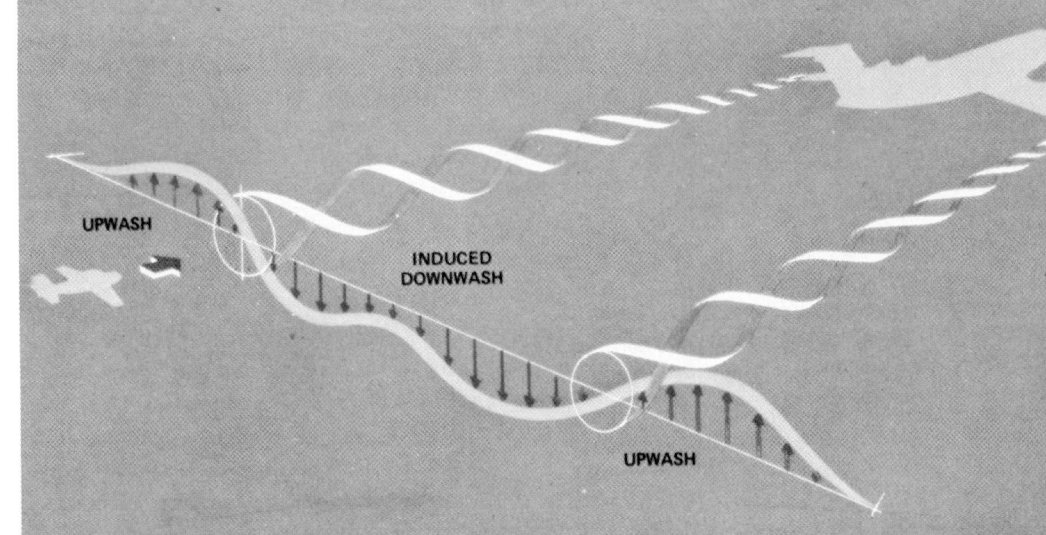

141 Pattern of turbulence caused by wing-tip vortices.

3-D RNAV system, or by reference to a ground scanning-beam microwave landing system.

Recognizing that noise-abatement procedures by themselves are not sufficient to bring aircraft noise down to acceptable levels, regulations in the United States impose noise limitations on future engine designs, and in some instances call for a retrofit of existing engines.

Noise is measured in "effective perceived noise in decibels" or "EPNdB" which are computed by a mathematical formula that takes into consideration the effect on the ear of tonal strength and duration of noise, as well as loudness and frequency. Aircraft noise under the United States regulations is measured at one nautical mile from runway threshold for arriving aircraft; and in the case of departing aircraft, 3.5 miles from start of takeoff roll, as well as "sideline" noise on both sides of the departure path. The prescribed noise limits eventually will reduce aircraft noise by as much as 10 EPNdB. A reduction of 10 EPNdB represents a 50 percent reduction of perceived noise.

WAKE TURBULENCE

Every aircraft generates a pair of counter-rotating vortices trailing from the wing tips causing "wake turbulence." As aircraft become larger and heavier, the intensity of the vortices can pose problems for smaller aircraft. In fact, large aircraft generate vortices with roll velocities exceeding the roll-control capability of some smaller aircraft. Wake turbulence is not the result of the exhaust thrust of jet engines.

Vortex strength depends mostly on the weight, speed, and shape of the wing of the generating aircraft. Vortex characteristics can also be changed by extension of flaps and other wing-configuring devices as well as by a change in speed. The basic factor is weight, however, and the vortex strength increases with increased weight and span loading. Vortex tangential velocities can be 150 feet per second or about 90 knots. The greatest vortex strength occurs when the generating aircraft is heavy, clean (no flaps, spoilers), and slow.

As a consequence, wake turbulence is a factor which affects the spacing which ATC applies between aircraft en route, landing, and taking off. Since this spacing in some situations is greater than the normal spacing required for traffic-separation purposes, runway/airport capacity as a result is further reduced during the operation of certain types of aircraft.

Heavy Jets

Based on extensive tests to detect and measure the effects of wake turbulence, the United States — and most other countries — categorizes all aircraft capable of takeoff weights greater than 300,000 pounds as "heavy jets." This category includes such aircraft as the Boeing 747, Douglas DC-10, the Lockheed L-1011 and C-5A, and perhaps the European "Airbus." When in flight, the vortices sink downward and decay. On landing, they stop after touchdown; on departure, they commence after rotation. The strength and direction of the wind affect the course of the vortices until dissipation.

Separation Minima

Heavy jets operating behind heavy jets generally do not require any special separation because of wake turbulence. In the case of all other aircraft, however, "wake-turbulence" separation minima are imposed by ATC. Although quite extensive, the following examples illustrate these minima.

- When only radar separation is being applied, a minimum of 5 miles is provided between a heavy jet and any other IFR aircraft operating directly behind it, i.e., in the six o'clock position.
- When a VFR aircraft is being radar vectored or sequenced behind a heavy jet, i.e., at the six o'clock position, a minimum of 5 miles is provided unless the VFR aircraft is known to be above the heavy jet or 1,000 feet or more below it.
- At least a two-minute interval behind a departing IFR or VFR heavy jet is applied to aircraft departing on a parallel runway separated by 2,500 feet or more if the succeeding departure flight path converges with the departure flight path of the heavy jet.
- At least a two-minute interval behind a VFR or IFR departing heavy jet is applied to aircraft landing on a crossing runway if the arrival flight path will cross the takeoff

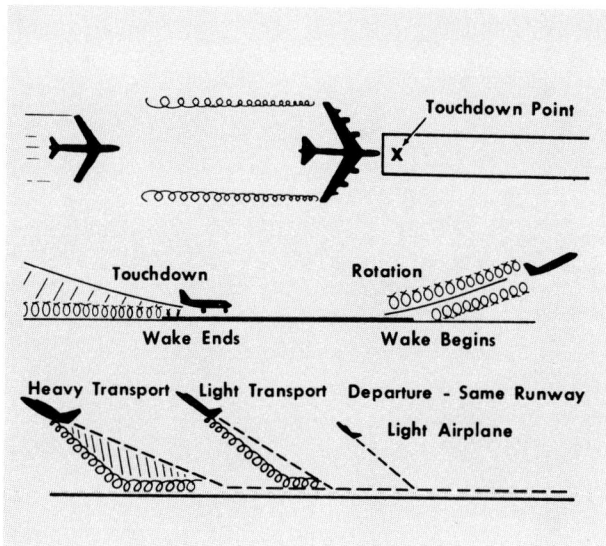

142 *Basic effects of wake turbulence.*

path behind the heavy jet and in front of the heavy jet rotation point.

• At least a two-minute interval behind a departing IFR or VFR heavy jet is applied to aircraft departing on the same runway, parallel runways separated by less than 2,500 feet, or crossing runways if projected flight paths will cross.

The net result of the application of wake-turbulence separation minima — the lighter the following aircraft, the greater the separation will be — is to reduce the normal CTOL runway acceptance rate, already at only about 50 percent of its maximum theoretical capability as discussed earlier, by another 50 percent. Thus, the productivity of a CTOL runway on which wake-turbulence separation minima are applied will be about 25 percent of its potential productivity.

Sensing and Suppression

Since the operational hazards and economic penalties resulting from wake turbulence are very severe, it is evident that elimination or at least significant reduction of wake turbulence is extremely important. The basic solution lies in remedial aerodynamic design. Secondly, sensing and dissipation devices at airports can supply some degree of relief. But V/STOL's offer another solution by virtue of their independent operational capability.

TERMINAL AIRSPACE

The airspace surrounding a principal airport or "terminal" is referred to as a "terminal area." A terminal airport generally is identified in relation to CTOL traffic density. The radius of a terminal area varies from 20 to 60 miles of the terminal airport. The vertical dimensions also are variable starting at the surface and extending upward to as high as 10,000 feet MSL. Within the terminal area there usually are several satellite airports in addition to the terminal airport. Traffic within the terminal area is controlled by the tower at the terminal airport through its approach and departure control facilities.

Basic Terminal Area

In a basic terminal area, aircraft may be flown in accordance with standard procedures depending upon whether under VFR or IFR. All terminal areas have radar facilities.

Each terminal area has a number of "transition points" marking the entry/departure routes for flight within the area. If an IFR aircraft is not equipped with area navigation (RNAV), extensive radar vectoring by the controller may be involved. If RNAV equipped, radar vectoring within the terminal area is minimal.

Standard STAR's and SID's, and RNAV STAR's and SID's are called for by the approach/departure controllers as appropriate to expedite traffic flow and reduce controller/pilot communications.

143 *Vertical cross-section and plan view of terminal area model.*

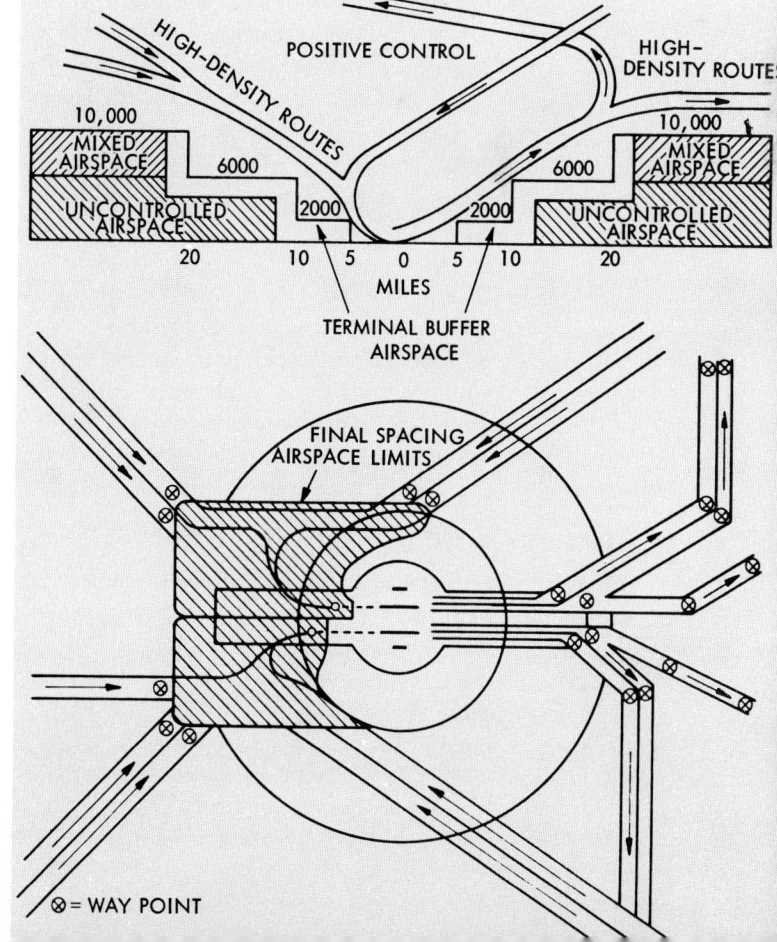

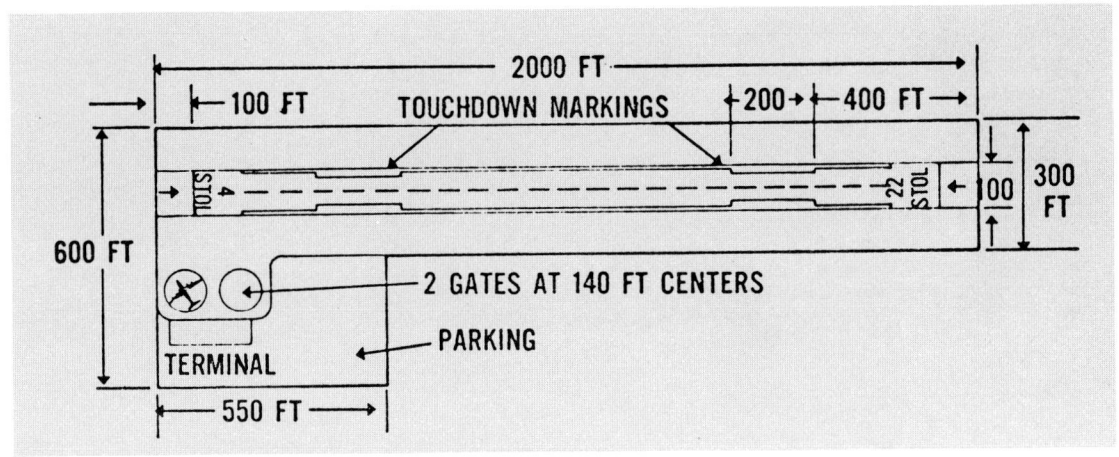

144 Minimum design for an independently located STOLport.

Terminal Control Area

A terminal control area, or "TCA," is established at selected terminal airports where the density and/or complexity of traffic within the area warrants. The outside horizontal and vertical dimensions vary in a manner similar to the standard terminal area. The TCA, however, defines a series of increasing levels from the surface to its ceiling wherein *all* air traffic, whether operating in accordance with IFR or VFR, is controlled by the terminal airport ATC facility.

To operate in a TCA, pilots are required to obtain an ATC clearance and follow ATC instructions. Sequencing and radar-separation service are provided by the ATC facility in accordance with IFR separation standards regardless of weather conditions.

Minimum-airborne equipment requirements include VOR or TACAN navigation equipment, two-way radio communications equipment, and a radar transponder (beacon).

AIRPORT GROUND-TRAFFIC CONTROL

The continuing progress toward eventual all-weather operations — day and night — focuses attention on the need for improved airport ground-traffic control. The ability of the tower controllers to visually supervise traffic movements on the runways, taxiways, and ramps is severely limited by the size of the airport, blind spots, and reduced visibility during inclement weather and at night. The pilot, on the other hand, during poor weather with little or no visibility, does not have positive guidance to assist him in staying on the runways and taxiways. Nor does he have the capability to detect other aircraft or vehicles on the airport's surface which may cause a collision hazard.

An airport ground-traffic control (AGTC) system must be designed so as to serve effectively *both* controller and pilot in all-weather conditions, IFR and VFR, day and night. Such a system performs two functions: surveillance, for the controller; and guidance, for the pilot.

Surveillance

Some limited but not very effective surveillance of an airport's surface is possible with the airport surface-detection radar equipment (ASDE) previously described. This type of equipment, however, has severe technical limitations and does not constitute an effective controller/pilot, airport ground-traffic control *system.*

Basically, the surveillance subsystem of a properly designed AGTC system provides the controller with:

- Information on the identification, position, and progress of each aircraft at all times.
- Information on the position and progress of essential ground traffic at all times.
- Information on the presence of obstructions and temporary hazards.
- Information on the operational status of the system in use.

The system must assure ground vehicles of:
- Adequate routing, navigation information, and means of collision avoidance to permit rapid access to any part of the airport, and, also, if practicable and economic, outside the perimeter. Crash-location equipment must also be considered.
- Safe and positive routing as they fulfill their roles of attending to the essential needs of aircraft passengers and airline and airport personnel and services.

Guidance

The guidance subsystem provides the pilot with:
- Steering and distance information in order to track along the runway, decelerate, and reach the turnoff point safely and efficiently on landing; conversely, on takeoff.
- Visual display of information as to the route to be followed to/from loading/unloading position.
- Assurance of separation from other aircraft and vehicles on the airport's surface.
- Adequate warning for changes of direction or the need for speed adjustment.

Subsystem Interaction

The surveillance and guidance subsystems interact to:
- Provide an autonomous aircraft intersection/crossing control system to relieve the controller from surface-conflict detection and resolution tasks.
- Relieve routine control and routing responsibilities for the controller.

145 *Even in major metropolitan areas, there are many locations from which to choose a site for an independent STOLport.*

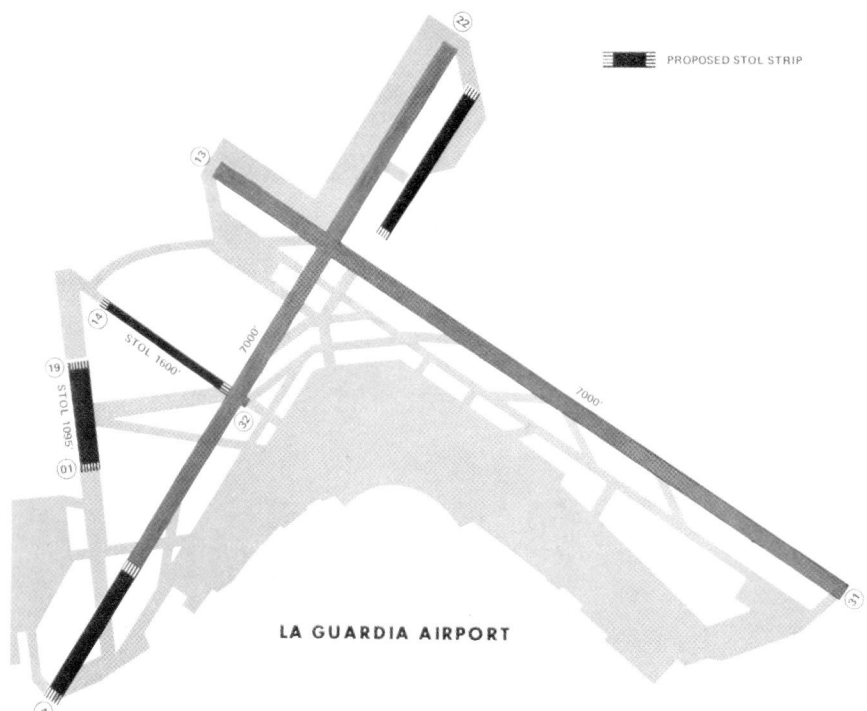

146 *Example of using a major CTOL airport to provide STOL strips by taking advantage of existing runways and taxiways.*

- Provide independent operation of one subsystem in the event of failure of the other.
- Reduce pilot/controller communication workload.
- In general, increase airport surface safety and expedite aircraft and other vehicular movements.

STOLPORTS

STOLports are of two basic types.
- CTOL STOLports, consisting of one or more STOL runways situated at a CTOL airport.
- Independent STOLports, consisting of one or more STOL runways situated at a site removed from and independent of a CTOL airport. This type of STOLport is subdivided into two further categories.

  Surface STOLport, in which the STOL runway(s) are located on the surface.

  Elevated STOLport, in which the STOL runway(s) are located on an elevated structure of some sort.

Basic Design Criteria

Certain basic design criteria are applicable to all STOLports, regardless of type. These criteria relate essentially to safety, noise, environment, and economics.

Ideal configuration for a STOL runway at sea level with 90° F temperature is 1,500-1,800 feet. A minimum of 100 feet protection surface is required at each end, thus making the typical STOL runway about 2,000 feet in length. Typical STOL runway width is a minimum of 100 feet and a maximum of 300 feet.

STOL runway orientation is influenced primarily by the wind factor. The runway should be aligned with the prevailing winds, although the limited number of STOLport sites available, particularly in metropolitan areas, may minimize the opportunity for a particular runway to have optimum wind coverage. Also, availability of space for a cross-wind runway on a metropolitan STOLport will be rare. Nonetheless, the design objective is to attain at least 95 percent and preferably +98 percent usability of a STOLport in all wind conditions.

Where traffic volume and space permit, parallel runways may be provided at a STOLport. These may be spaced as close as 700 feet center-line to center-line, which is considered sufficient spacing for simultaneous VFR operations. Separation between center-lines of parallel runways for simultaneous IFR operation will depend upon navigation accuracy, aircraft performance, and capability of the Air Traffic Control System. The desired center-line separation goal for simultaneous IFR operations is somewhere between 2,500 feet and 1,000 feet.

STOL runways at a CTOL airport may be a portion of an existing runway specifically marked for STOL use under certain weather conditions, or a special STOL runway may be constructed at some suitable location within the airport area. Separation between STOL runways and CTOL runways to permit simultaneous STOL and CTOL operations in both VFR and IFR conditions requires thorough analysis in each instance. ATC separation procedures,

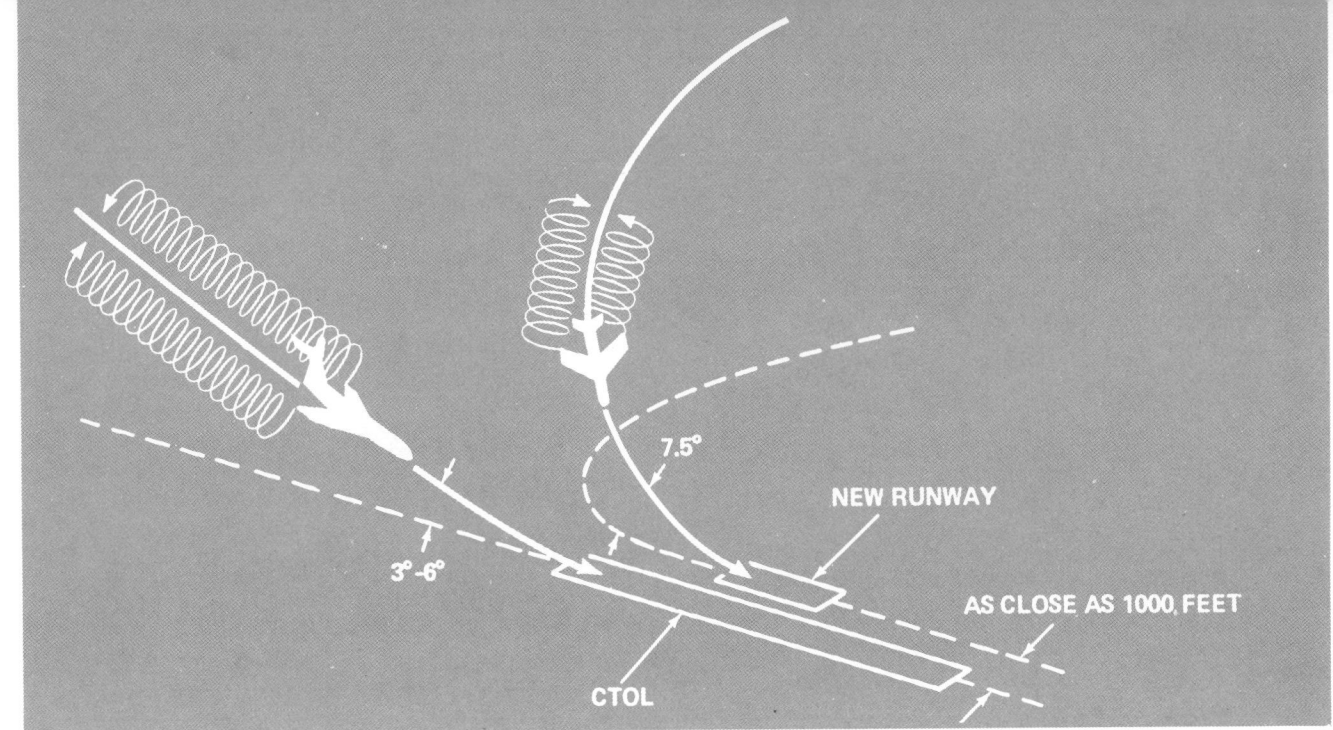

148 Mixed CTOL and STOL traffic operating independently on closely spaced parallel runways with no wake turbulence interaction.

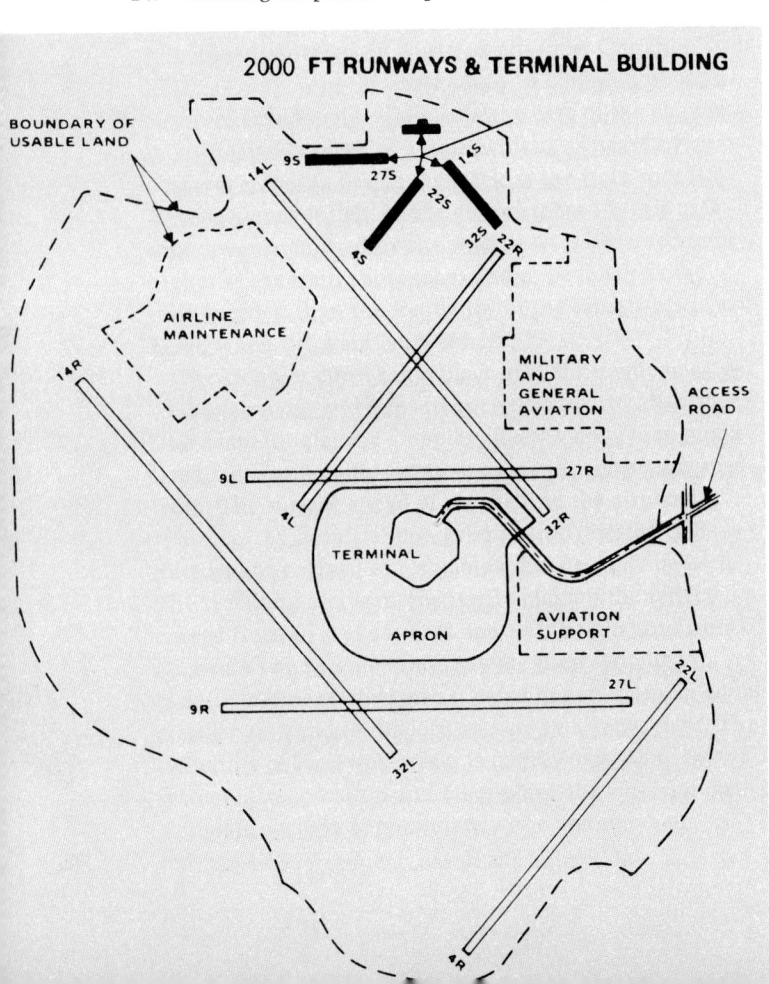

147 Creating a separate STOLport area on a major CTOL airport.

CTOL traffic patterns, and other ATC factors must be taken into consideration in order to permit the simultaneous operation of the two different types of air traffic to and from the same airport, but without causing interference with each other.

STOL Approach/Departure Profiles

The STOL aircraft offers a distinct advantage over the CTOL aircraft in laying out STOLports due to the inherent characteristics of the STOL aircraft to descend and climb at steep gradients. For example, a STOL aircraft can make an approach to the touchdown zone (TDZ) on a runway at a descent gradient of up to 8.5°, whereas the normal descent gradient for approach to a landing by a CTOL aircraft is 3°. Also, the STOL aircraft can follow an equally steep or even steeper ascent gradient.

Furthermore, the characteristics of the STOL aircraft make feasible a *curving* approach (or departure) in both VFR and IFR—assuming suitable navigation facilities—which can provide considerable benefits to and flexibility in the ATC System. For example, an arriving STOL aircraft after entering the terminal area can proceed on a discrete three-dimensional routing to its initial approach fix. Thereafter, the STOL can make a curving approach going downwind, parallel to the STOL runway, at the same time descending at a steep (7.5°) gradient to touchdown point for a landing. With this type of procedure, the STOL aircraft may make a simultaneous IFR approach with a CTOL aircraft landing on a closely spaced, parallel CTOL runway. The characteristics of STOL aircraft will permit this type of curving approach on a radius of about 1,500 feet.

In general, STOL operations to and from a STOLport offer a great deal of flexibility in providing discrete routings to facilitate traffic flow. Airspace required by STOL aircraft for takeoff and landing and in terminal area maneuvering is significantly less than that required by CTOL aircraft. These characteristics not only facilitate operation of the ATC System in segregating STOL and CTOL traffic, but also assist in carrying out obstruction-clearance and noise-abatement procedures, including two-segment approaches as appropriate. The curving, steep gradient, flight paths of STOL aircraft can be so arranged as to avoid any adverse effects of wake turbulence from heavy CTOL aircraft.

Other Considerations

Other factors involved in STOLport design and operation affect the ATC System directly or indirectly. These include such considerations as the layout of the taxiways to facilitate rapid access to and egress from the STOL runway, as well as to and from loading/unloading areas. Essential to all-weather operation, one or more runways on the STOLport (or STOL runways on a CTOL airport) must have a suitable precision-landing system. If traffic volume is sufficiently high (same as CTOL criteria), a control tower will also be needed.

Other STOLport requirements include emergency arrest-

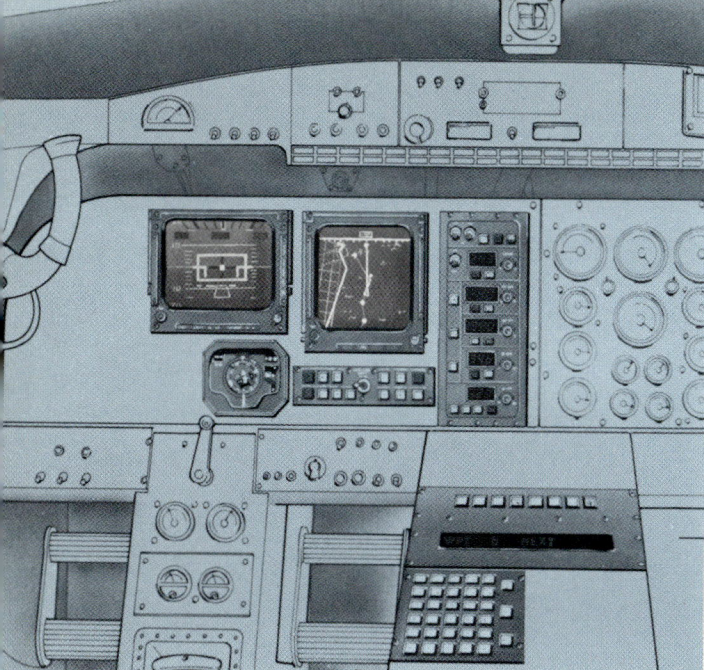

149 STOL navigation and landing system (STOLAND) incorporates automatic guidance and landing capability as well as 4-D RNAV.

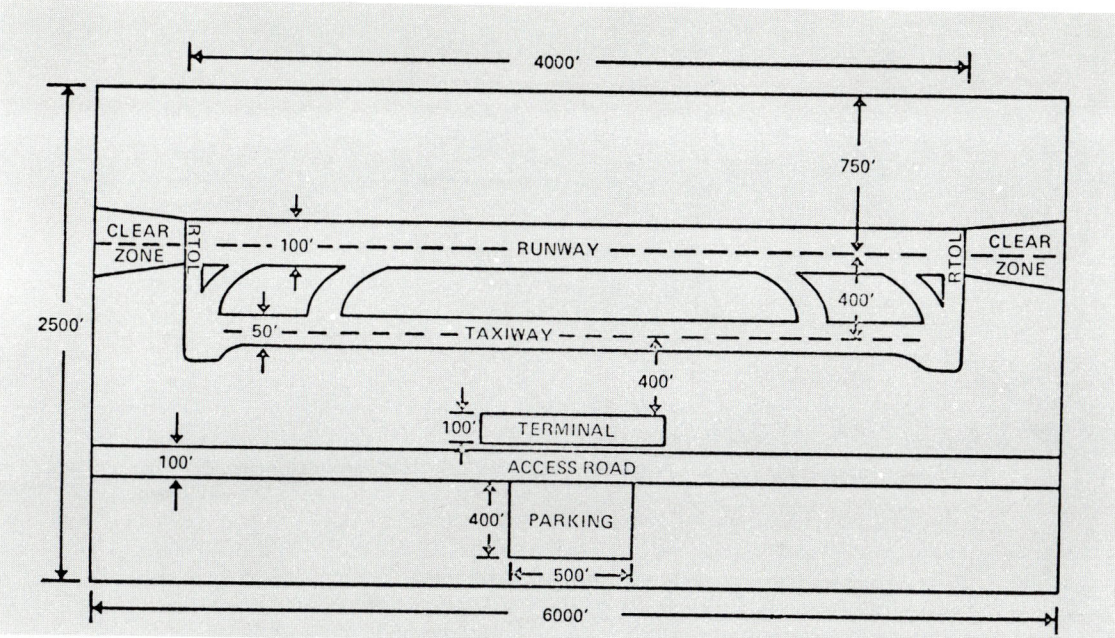

150 Basic reduced takeoff/landing (RTOL) facility.

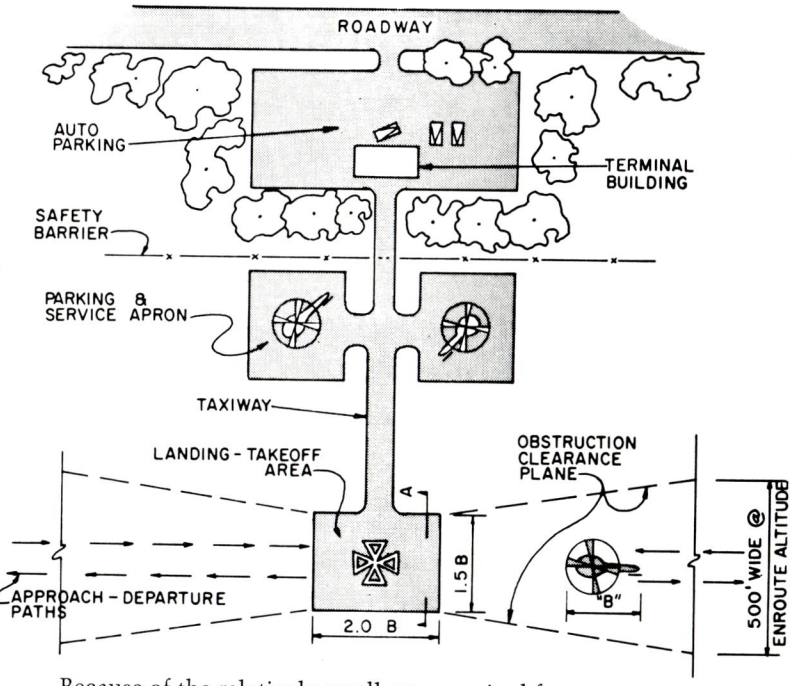

151 Surface VTOLport model for VTOL aircraft.

ing systems or barriers (especially important for elevated STOLports); emergency equipment such as fire-fighting systems; runway marking (same as for CTOL airports except ends are marked "STOL"). Also required is an adequate runway lighting system. This is basically the same as CTOL runway lighting except for the visual-approach slope indicator (VASI) which is set for a glide-path angle of 6° to 8°, and a special high-intensity STOLport beacon.

RTOLPORTS

All of the considerations applicable to STOLports are applicable to RTOLports with one principal difference—the runway length. In the case of an RTOL aircraft, a runway length of 4,000 feet is required versus the STOL's 2,000-foot runway length. An aircraft designed as a STOL, however, may in certain situations be able to operate with a greater payload into and out of an RTOLport than that possible at a STOLport. Another difference is that elevated RTOLports are not considered feasible.

VTOLPORTS

As with STOLports, VTOLports may be located within the boundaries of a CTOL airport or they may be at an independently located site. Similarly VTOLports may be located on the surface as well as on elevated structures such as the tops of buildings. VTOL aircraft also may use a STOLport or RTOLport in mixed STOL/RTOL traffic because of their somewhat related performance characteristics.

Basic Design Criteria

Because STOLports require a significant amount of area, their number of necessity must be somewhat limited. On the other hand, VTOLports, with their much smaller space requirements, can reach a tremendous volume. If unprepared VTOLports, such as "helipads," are considered, the potential number of VTOLports can be virtually unlimited. This latter situation would be particularly true with the advent of widespread general aviation VTOL operations.

Since VTOL's require little or no takeoff and landing roll, the size of the landing/takeoff area is largely influenced by the size of the VTOL aircraft which will use it and the overall size of the VTOLport by the traffic volume to be accommodated.

To postulate, a large airline-type commuter VTOLport with dimensions of 300 feet by 450 feet could have four gate positions of 150 feet by 150 feet each, plus a lift-off/touchdown area in the middle of 150 feet by 300 feet. With multidoors for quick passenger exit/entrance, on a VTOL carrying 100 passengers, loading and unloading times could average perhaps 6 minutes, or 10 operations per gate per hour. A commuter VTOLport in this category could thus handle in the order of 40 landings and 40 takeoffs per hour (i.e., one operation every 45 seconds) with up to 8,000 passengers (in and out) per hour. Such a VTOLport, as in the case of STOLports, would be served by supporting ground transportation and other pertinent facilities.

Because of the relatively small area required for VTOLports, they may be constructed more readily in city centers and urban areas than is the case with STOLports. Potential sites include park areas, piers, and tops of buildings. Because of their location flexibility, VTOLports can provide many opportunities for expanded air transportation services not possible with STOLports and CTOL airports.

152 Elevated VTOLport located on top of a major building.
153 VTOL commuter air-bus concept.

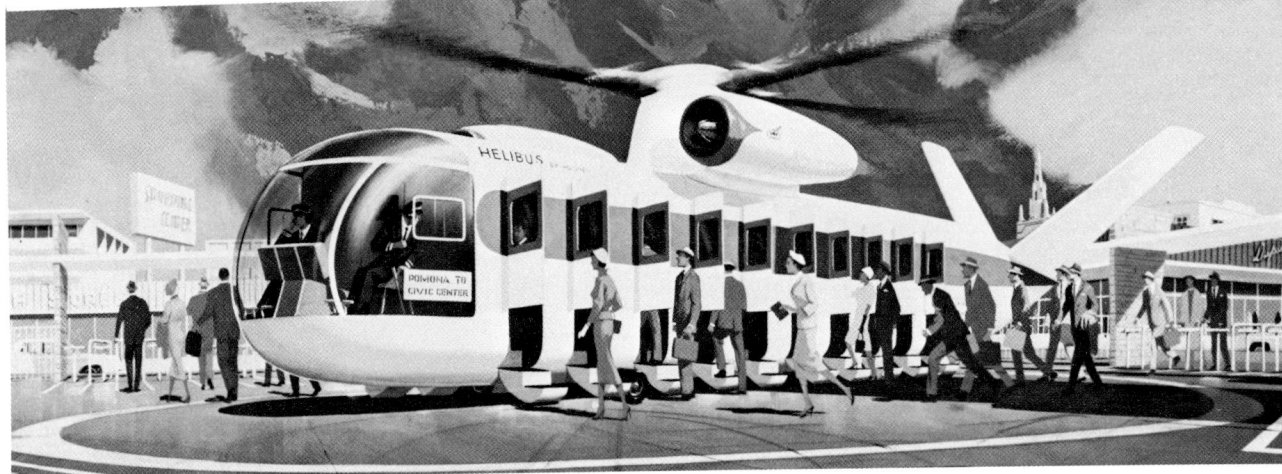

154 Commuter VTOLport and STOLport over surface transportation convergence.

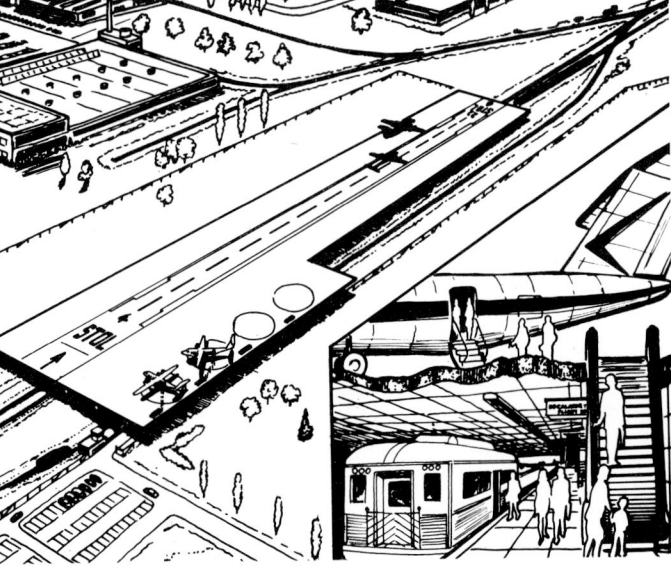

As in the case of elevated STOLports, elevated VTOLports may cause an increased incidence in instrument approaches over those required at surface STOL and VTOLports. With a 300-foot ceiling, for example, a surface facility would have that same ceiling, but an elevated facility would have a relatively lower ceiling depending upon its height. If the VTOLport was on a structure 300 feet high, a Category III C instrument approach might be involved, whereas at the surface facility only a Category I approach would be needed.

Although an elevated VTOLport as well as an elevated STOLport have some disadvantages, such as a higher incidence of instrument approaches, they also have some advantages. Included are less obstacle-clearance hazards, reduced exposure to turbulence and buffeting, and lowered noise exposure on the surface.

A particular advantage of the VTOLport over the STOLport and CTOL airport is that there is no wind orientation problem. The VTOL aircraft can follow the prescribed approach path to the VTOLport and then at point of hover turn into the wind prior to touchdown.

VTOL Approach/Departure Profiles

From an Air Traffic Control standpoint, a VTOL aircraft has all of the STOL's advantages of flexible flight path and variable ascent/descent gradients, plus some significant additional advantages. These include a greater range of variable descent gradients ("normal" 6° to 15°, "steep"

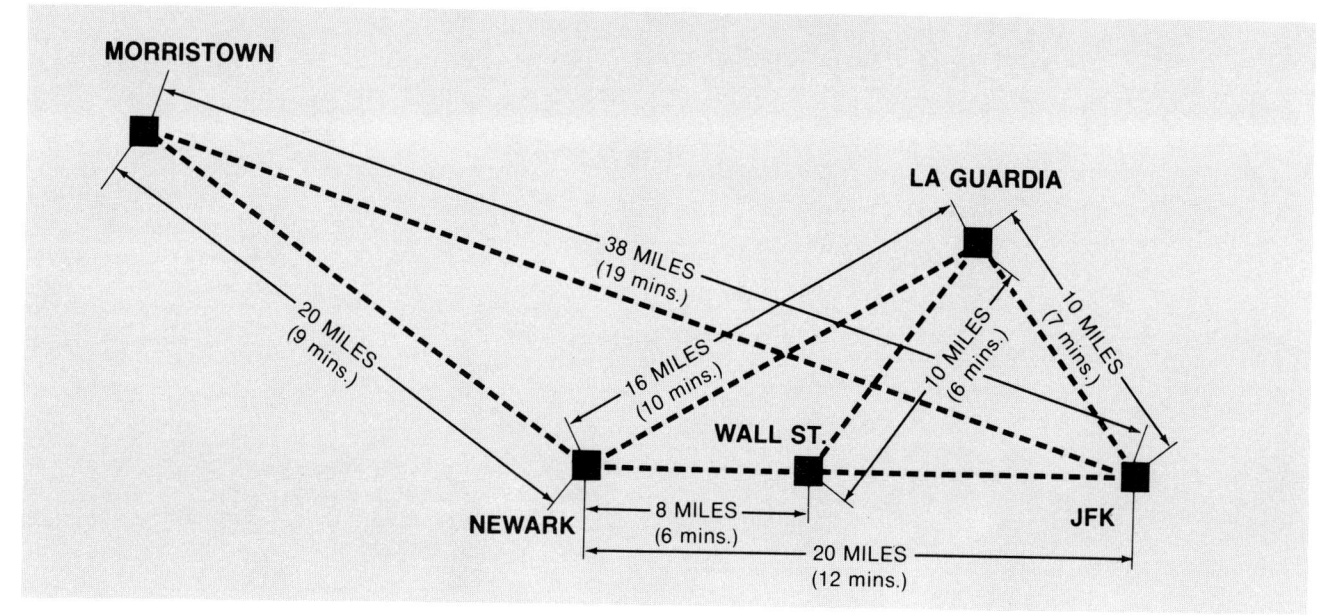

155 VTOL service on scheduled basis in an intraurban area involves short segments and precision routings, with attendant requirements for highly efficient air traffic control.

15° to 30°, and "vertical" 30° to 90°); steep climb gradients; ability to quickly slow down speed and, if necessary, stop in midair; and ability to maneuver within very small amounts of airspace.

With the VTOL's capability to achieve four-dimensional—lateral, longitudinal, vertical, and time—flight profiles in an extremely accurate manner, the ATC System can greatly increase the capacity of airspace and airports. But to fully take advantage of these capabilities, as well as those of the STOL, highly accurate airborne and ground-navigation equipment and other system improvements are needed.

Other Considerations

ATC related requirements at an airline or commercial type VTOLport are about the same as for a STOLport insofar as local navigation and tower facilities are concerned. The instrument landing equipment, however, will need to be designed in order to accommodate the VTOL's wide range of variable descent gradients. Distinctive VTOLport markings are used; emergency equipment is needed. VTOLport lighting requirements are distinct from those of STOLports and CTOL airports since there are no fixed approach paths nor runways/taxiways. As a consequence, the entire VTOLport may be floodlighted and other lighting innovations will be needed for all-weather VTOL air service.

156 Model VTOL operation following preprogrammed 3-D area navigation profiles, combined with local precision approach facilities, provide basis for independent VTOL/CTOL traffic flow.

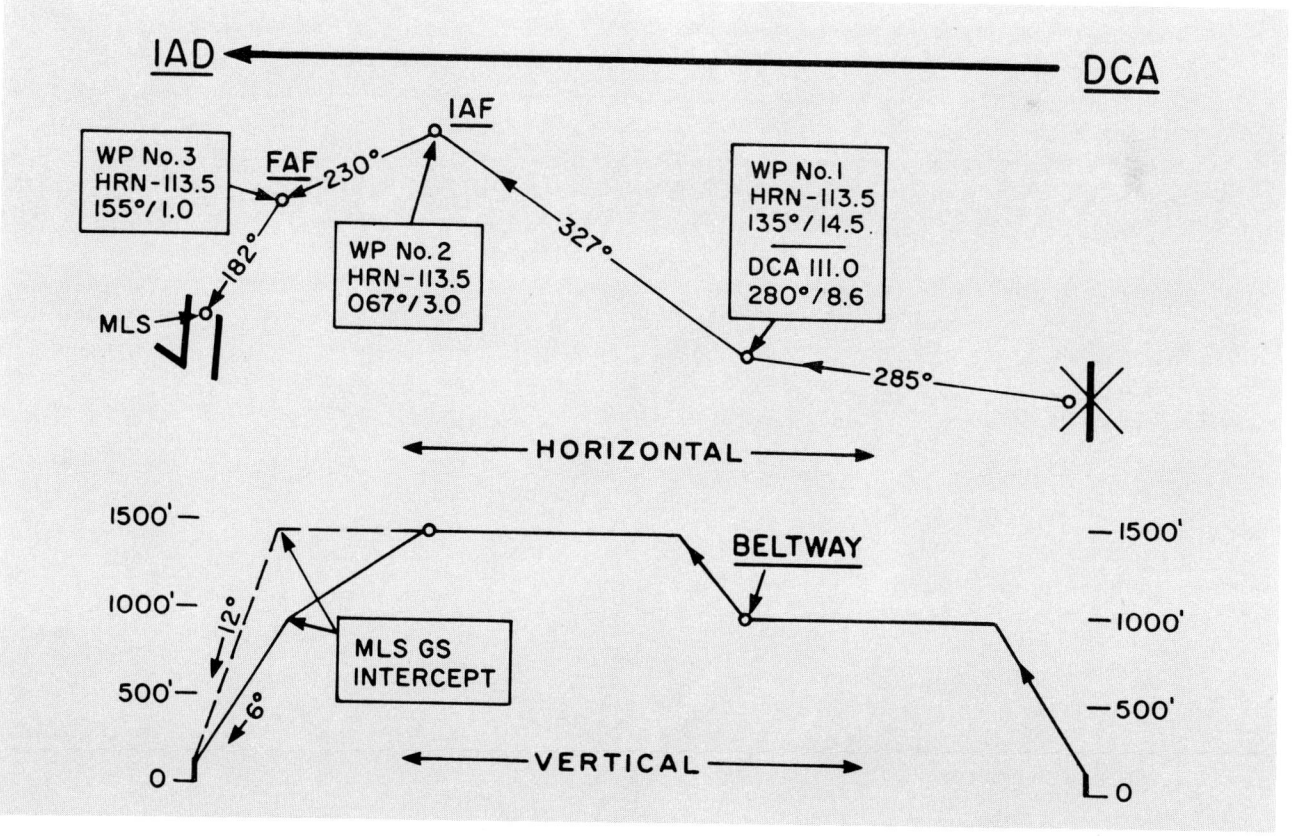

The productivity of a major airport can be maximized by providing dedicated facilities to serve the different classes of user. Discrete routings to and from the separate facilities are followed by the pilots to facilitate the functions of air traffic control in achieving maximum airport efficiency and productivity.

CHAPTER XI

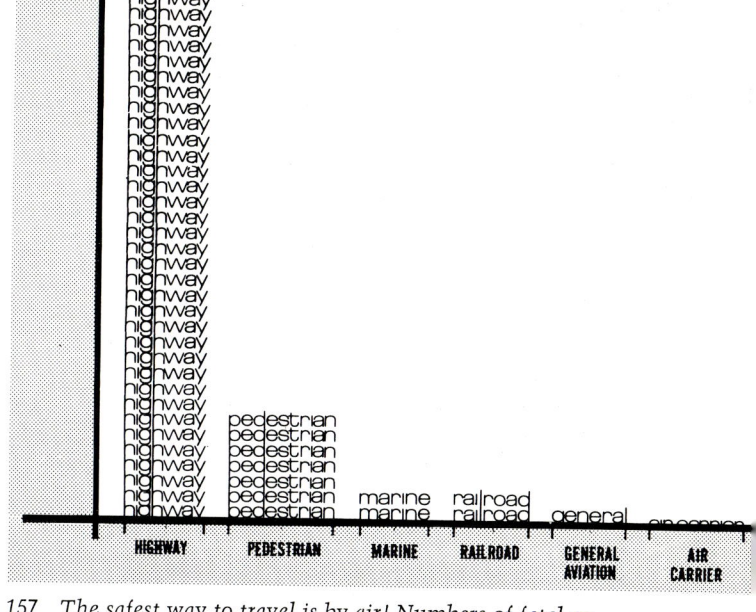

157 *The safest way to travel is by air! Numbers of fatal accidents are far less when flying than when using surface methods of transportation. Nevertheless, air safety must constantly be raised; preventing midair collisions is an essential part of this program.*

COLLISION AVOIDANCE

The avoidance of collisions between aircraft—the primary objective of the Air Traffic Control System—depends upon two basic sources of air traffic information: *ground derived;* and *air derived.*

In some instances, only one or the other of these information sources is available to the pilot or controller. In others, both may be used in combination to assist on the decision-making process to avoid collisions with other aircraft.

Collision avoidance based on ground-derived information relies on the ability of controllers to detect a potential collision situation in advance and instruct the pilot as to an appropriate avoidance maneuver. Air-derived collision-avoidance information uses either or both of the following techniques.

• Visual detection by the pilot of other conflicting traffic sufficiently in advance to take avoiding maneuvers.

• Instrumental detection of other potentially conflicting air traffic, giving the pilot either: cues as to approximately where he should look in order to detect such traffic visually; or command indications as to the avoiding maneuver which he should take.

THE COLLISION ENVIRONMENT

Based on an analysis of midair collision data gathered in the United States over a significant period of time, the environments in which collisions between aircraft are most likely to take place can be identified. These conclusions are based not only on actual midair collisions, but also on an extensive sampling of "near" midair collisions as reported to the Federal Aviation Administration by pilots. Near midair collisions are defined as incidents in which a potential midair collision was averted only by chance, and those in which a probable collision was prevented by one or both pilots taking avoiding maneuvers.

The two broad environments in which midair collisions take place are the terminal area—within 30 miles of an airport—and en route. The terminal area provides the environment in which a majority of collisions occur.

Within the terminal area, about 75 percent occur within 10 miles of the airport. Of this percentage, approximately 85 percent take place at an impact altitude below 5,000 feet above ground level (AGL), and around 65 percent below an impact altitude of 3,000 feet AGL. Most of these collisions involve aircraft on overtaking convergence courses and flying in *good weather.*

En route, the environment in which the midair collision exposure is most prevalent is within the vicinity of a VOR ground station due to the concentration of aircraft over these facilities when flying the VOR-to-VOR "airway" system. Widespread use of "area" navigation reduces this incidence factor.

As between the different classes of airspace users, distribution of midair collisions may be expected to be somewhat as follows:

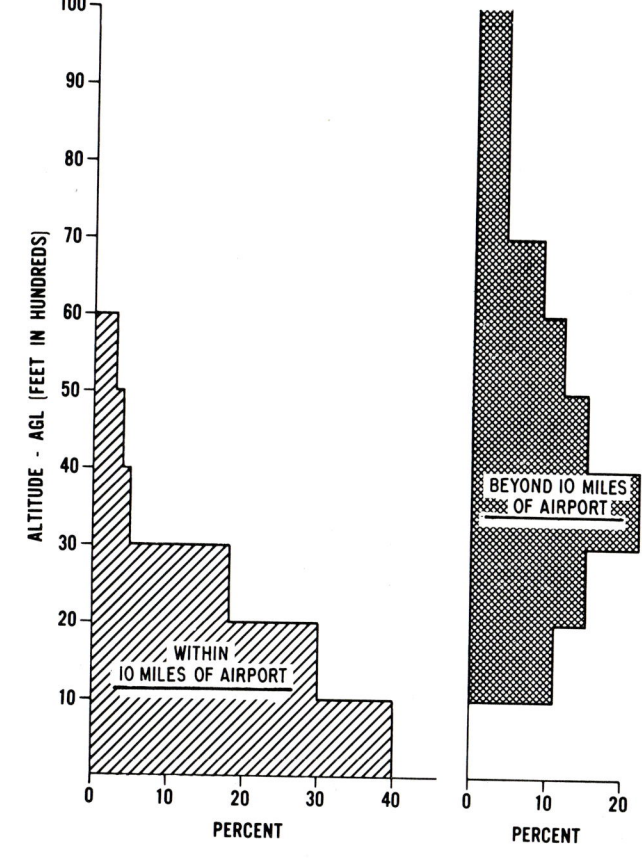

158 *Midair collision exposure related to distance from airport and altitude.*

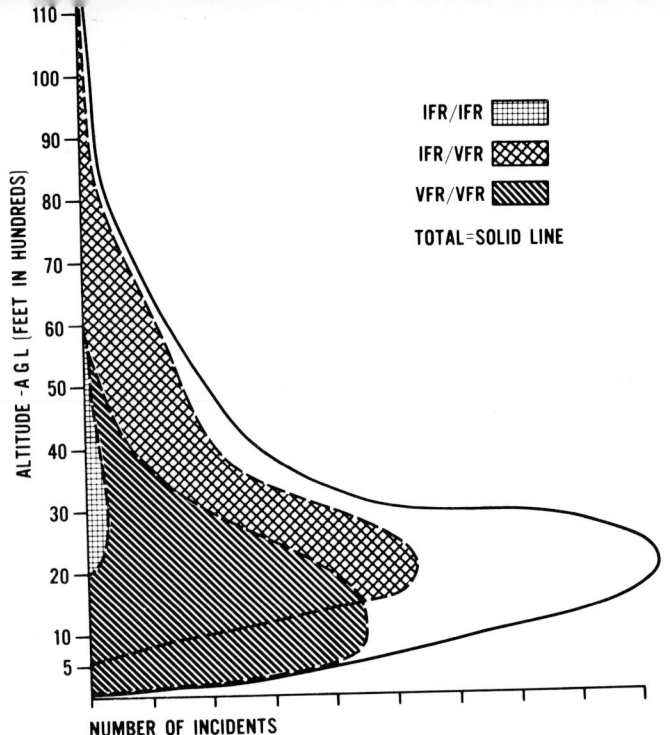

159 Distribution of midair collisions by type of flight and altitude.

160 The controller plays a key role in the ground-based collision-avoidance ATC System.

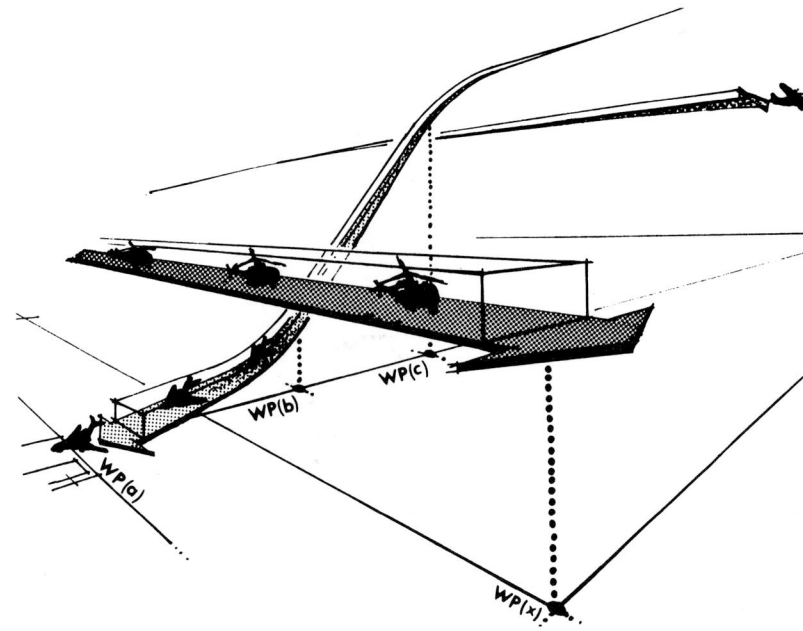

161 Airspace is structured in three dimensions, using RNAV waypoints as appropriate, to assist in the collision-avoidance process.

	Percentage
Air Carrier/Air Carrier	— 1
Air Carrier/General Aviation	— 4
Air Carrier/Military	— 2
General Aviation/Military	—10
General Aviation/General Aviation	—83

In summary, the environment most favoring midair collisions is that within 10 miles of an airport, below 5,000 feet AGL and during VFR weather. The next most likely environment for midair collisions is in the vicinity of a VOR station during en route flight. Collisions between general aviation aircraft provide a major challenge to ATC System engineering.

GROUND-DERIVED COLLISION AVOIDANCE

The primary tool used by the controller to provide separation assurance between aircraft and thus avoid collisions is through the use of radar surveillance, as described previously. If the radar system is associated with a sophisticated computer complex (alphanumerics), the controller is assisted in his collision-avoidance decision-making process by means of computerized conflict detection and resolution displays.

In effect, this provides a ground-based collision-avoidance system (CAS). When flying under this system, the pilot is given continuous, detailed instructions by the controller as to flight parameters to be observed in order to avoid a collision with other aircraft. When all aircraft in a given environment are operating under full control of the ground-based CAS, statistics show an extremely low rate of midair collision incidence; however, the extent to which this method of collision avoidance can be expanded to handle constantly increasing traffic volume appears to be limited.

In mixed airspace where some aircraft are under the control of the ground-based CAS and others are not (i.e., some on IFR flight plans and others on VFR flight plans or on no flight plans), the midair collision rate rises in proportion to the admixture of controlled and uncontrolled traffic. To minimize this situation, certain portions of the airspace are designated as "positive-control areas" (PCA) in which all aircraft must be under the control of the ATC System at all times regardless of weather.

Another method of helping to alleviate the potential collision risk in a given environment is for the establishment by the ground system of traffic segregation rules. These may include separate controlled routings based on aircraft performance, e.g., high performance; separate routings for controlled and uncontrolled operations, e.g., "VFR freeways"; and discrete routings for different aircraft types, e.g., CTOL, STOL, VTOL.

In the uncontrolled airspace, the midair collision rate is at its highest. It is in this environment that the greatest percentage of midair collisions takes place, that is, those between general-aviation aircraft. It also is in this environment that the ground-based CAS is ineffectual.

AIR-DERIVED COLLISION AVOIDANCE

This area of collision avoidance presents a number of aspects which basically affect the overall collision-avoidance capability of the ATC System. Developments in air-derived, collision-avoidance techniques will determine the difference between an ATC System which makes possible the use of airspace and airports both safely *and* efficiently, or one which restricts airspace/airport use in the interest of preventing collisions between aircraft at the price of minimizing the use of the airspace as an effective transportation medium.

Basic Considerations

The flow of aircraft in the airspace is not unlike that of surface automobile traffic flow in many basic respects. Both in the sky and on the surface there are superhighways, overpasses and underpasses, country roads, one-way routings, minimum (as well as maximum) speed zones, and other comparable situations over which in each case an overall traffic control system is imposed with the objective of preventing collisions.

But there are some basic differences. An automobile operates in two dimensions within confined routes on the surface; an aircraft operates in three dimensions on any desired routing. An automobile driver avoids collision by visual surveillance of the movements of other vehicles. He has reasonably good visual range in all directions (with the aid of rear-view mirrors); he can slow

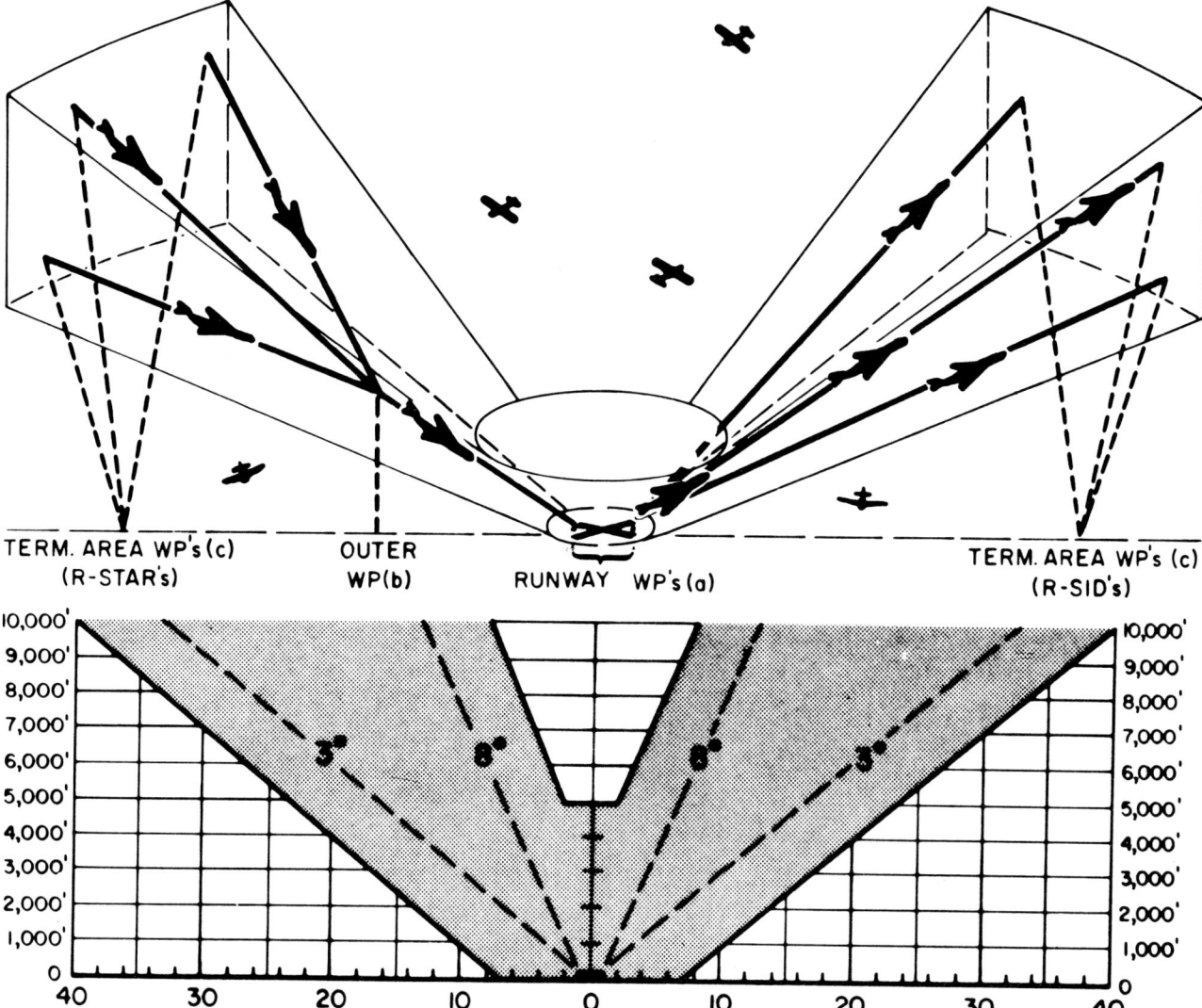

162 *Protected ascent/descent corridors can provide segregation between high-performance controlled aircraft and other air traffic, and thus contribute to reducing midair collision hazards.*

down or stop readily if visibility becomes restricted. The aircraft pilot also must rely to a large extent for collision avoidance on visual surveillance. But his vision is restricted to a relatively small wedge above and below him and to the right and left, the main visual range being more or less directly ahead. In addition, when in reduced or no visibility, the pilot must continue moving forward at a relatively high speed even though he has no visual capability for surveillance of other traffic.

Conspicuity and Vision

The most fundamental way by which the midair collision hazard can be reduced is to insure that aircraft are as conspicuous as possible. This involves the use of bright-colored paint on selected surfaces; however, maintenance costs tend to reduce acceptance by many aircraft operators. Navigation and flashing high-intensity lights when used in all-weather conditions, day and night, provide a definite enhancement of aircraft conspicuity.

Improved pilot-compartment external vision is equally fundamental. This involves increasing downward and upward vision angles, as well as side vision capability. Obstructions to vision, such as windshield posts, are undesirable. Provision is needed to reduce the visibility limiting effects of glare as well as to keep the windshied clear during precipitation conditions.

Different types of approaches for a landing constitute another factor which affects cockpit visibility. If the ap-

163 *Basic proximity warning instrument (PWI) gives pilot a rough idea—"one o'clock"—as to where another aircraft is so that "see and avoid" maneuvers may be employed as appropriate.*

proach is shallow (e.g., 3°), a CTOL aircraft's nose may be on or above the horizon with a resultant loss of visibility ahead and below. In a steeper approach, such as during the first segment of the two-segment noise-abatement procedure, the deck angle of the CTOL will be below the horizon, thus giving the crew enhanced visibility of the airspace ahead. On the other hand, V/STOL-type aircraft in a normal, steep-approach gradient will be in a level or slightly nose-up attitude, with attendant forward visibility restrictions.

Pilot Awareness

Also involved in reducing midair collisions by means of visual detection is the extent to which the pilot is aware of the midair collision potential and periodically scans the airspace within his range of vision. This awareness, however, may be mitigated by preoccupation with flying the aircraft. Cockpit duties, including ATC communications, may result in the pilot flying "head-down," particularly in the terminal area, and thus being unaware of surrounding air traffic. Other factors which may adversely affect pilot collision-hazard awareness include fatigue, lack of adequate training, unfamiliarity with segregated or other special traffic control collision-avoidance routings, or just plain disregard for the basic right-of-way rules.

Even with more than one pilot in the cockpit, collision awareness may be somewhat limited. For example, visual traffic spotting by a three-man airline crew of traffic in a terminal area under controlled test conditions (in good weather without ground ATC input) perhaps can locate about one-third of the aircraft within a 30° horizontal range straight ahead, but this awareness drops to less than 10 percent on either side.

Pilot alerting of possible collision hazards may be enhanced by "VFR traffic advisories" from an ATC facility. These are given only when controller workload permits, however, and are not intended to serve as a complete collision-avoidance service.

Another procedure used by the ATC System to reduce midair collision exposure in mixed traffic environments is to minimize the amount of time high performance (controlled) aircraft operate below 10,000 feet within a terminal area. In effect, arriving high-performance aircraft are kept at the highest possible altitude for as long as possible, and departing aircraft are climbed to the highest possible altitude as soon as possible after takeoff.

To assist the pilots in directly avoiding midair collisions, and as a backup or supplement to the ground-based, collision-avoidance system, a number of airborne instrumental detection methods may be employed.

AIRBORNE DETECTION INSTRUMENTATION

Two basic methods apply to the design of airborne-detection instrumentation which can provide the pilot with air-derived, collision-avoidance information.

- Noncooperative, in which a collision threat is determined without requiring that the intruder aircraft carry cooperating collision-avoidance equipment; and
- Cooperative, in which cooperating collision-avoidance equipment must be carried by the intruder in order that a potential collision threat may be determined by other appropriately equipped aircraft.

The cooperative method of providing air-derived, collision-avoidance information to the pilot is generally considered most feasible from a technical and economical viewpoint; however, some noncooperative, or self-contained, systems may be employed if their accuracy and other performance characteristics are compatible with cooperative systems.

Pilot Warning Instrument

A proximity warning instrument, or "PWI," is an airborne device which alerts a pilot to the presence of an intruder aircraft and supplies sufficient information to enable him to detect it visually. The pilot then assesses the threat and reacts as may be necessary to avoid a collision.

The primary application of the PWI is to assist the pilot in achieving safer separation between IFR/VFR and VFR/VFR traffic. These devices are not intended to provide for safer separation between IFR/IFR traffic under the jurisdiction of the ground-based ATC System.

Minimum performance capabilities of an effective PWI include the following.

164 Airborne collision-avoidance system (CAS) instrument gives pilot command instructions such as "fly up" (A), "fly down" (B), and "maintain level flight" (C).

- Detection of an intruder aircraft within a radius of 1 to 3 nautical miles.
- Display of the intruder's position within an azimuth coverage of 100° to the right and 100° to the left.
- Display of the intruder's altitude up to 1,500 feet above and below within the azimuthal and distance coverage.
- Being able to show at least three targets simultaneously.
- Minimal false alarms.

In effect, a PWI gives the pilot gross position information on other air traffic within his area of concern. This information is somewhat analagous to the VFR traffic advisories in which a controller may advise the pilot "traffic at 2 o'clock, westbound, below you." An added advantage with the PWI, however, is that both pilots presumably will detect each other's presence and thus greatly increase collision-avoidance probability.

Collision-Avoidance System

By general definition, the airborne collision-avoidance system, or "CAS," refers to any system using air-derived information which detects a potential collision hazard, calls attention of the pilot to the hazard, and displays the evasive action he should take to avoid a collision with the intruding aircraft. If connected to the autopilot, the CAS can give the command signals to alter the aircraft's flight profile as appropriate for collision avoidance.

An airborne collision-avoidance system gathers information about other CAS-equipped aircraft in nearby airspace, sorts out this information, and identifies those aircraft, if any, which pose a potential collision threat. To detect a potential collision, the equipment determines if an intruder is or will be at the same altitude and is flying a path that, unless changed, will produce a collision.

The data which a CAS uses include the following.
- Altitudes of intruder and protected aircraft.
- Rate of change of altitude for both aircraft.
- Range between intruder and protected aircraft.
- Rate of change of range between protected and other aircraft.
- Headings and airspeeds of other aircraft.

When one or more aircraft start a turn or vary their established rate of climb, descent, or level flight, the CAS equipment uses the range (distance) and range-rate (rate of closure) to predict the future flight path.

Once a potential collision hazard has been detected by the CAS, an indication is given to one or more of the concerned aircraft so that the pilot(s) may take nonconflicting action within appropriate time limits.

Incorrect or unnecessary evasive action signals—false alarms—can pose a serious problem to the pilot. A false alarm may be caused by improper functioning of the equipment or extra-sensitive receivers. Or, the CAS may evaluate as a potential collision hazard a perfectly safe traffic situation being regulated normally by the ground ATC System.

In high-density approach, landing, and departure traffic patterns, for example, many aircraft are in a constantly changing flight environment vertically, longitudinally, and laterally, but with necessary separation established by ATC. Yet many of these situations could fit CAS criteria as potential collision hazards at any given moment. This problem can be alleviated somewhat by providing two sets of criteria for collision prediction—one for "en route CAS" and one for "terminal area CAS." Alternatively, the CAS may be turned off while in a terminal area, but this is the environment where most midair collisions occur.

"Time Frequency" (T/F) technology provides the basis on which the airborne CAS functions. The T/F system refers to the use of airborne "clocks" (oscillators) that are periodically synchronized in such a manner that precise time and frequency references are established and maintained between all aircraft in a predetermined flight environment. Basically, this technique permits the determination of range and range-rate between aircraft within selected altitude levels. From these data, an airborne computer predicts a potential collision course between any two equipped aircraft. The "clock" oscillators in each aircraft are synchronized by means of radio transmissions with a ground radio station operating in a manner such as to maintain an accuracy of these clocks in the order of two-tenths of a millionth of a second.

In order for the CAS concept to be effective in collision-avoidance application, all categories of air traffic need to

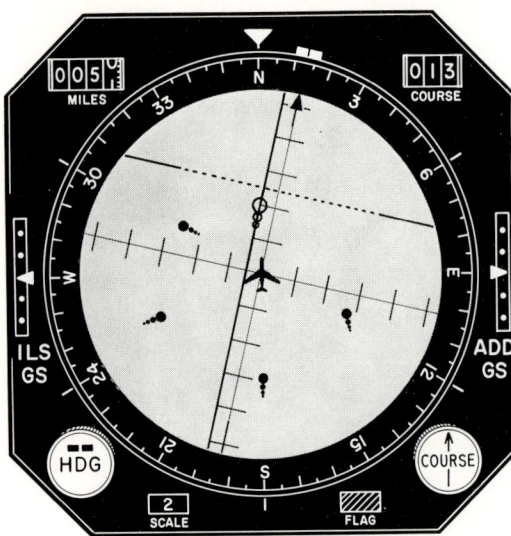

165 *In one concept of an air traffic situation display (ATSD), a cathode ray tube (CRT) instrument provides electronically generated navigation guidance combined with a background display of other air traffic within a pilot-selected, three-dimensional, safety envelope.*

166 *Another form of ATSD uses a CRT on which a moving pictorial display of navigation information is presented as well as positions of other aircraft in a selected area of concern.*

be equipped—airline, general aviation, and military. Economics and cost benefits constitute factors to be considered in achieving widespread implementation of the CAS. "Mini-CAS" equipment may be needed, as well as expanded application of the T/F technology to provide other functions.

Air Traffic Situation Display

In this category of airborne collision-avoidance instrumentation, the pilot is provided with a real world display of the traffic situation in his immediate area of concern, commonly referred to as an air traffic situation display—ATSD.

Based on such air-derived information, the pilot can assess a given traffic situation and maneuver his aircraft accordingly. Or, in cooperation or coordination with the ATC System, the pilot may carry out directly—by reference to the ATSD—certain functions to "unload" the ground ATC System, and at the same time increase airspace/airport capacity. These include:
- Station-keeping, which involves flying within specified close proximity to other aircraft.
- In-trail spacing at optimum distances from other aircraft departing, en route, and landing.
- Passing (going around) aircraft being overtaken.
- Avoiding crossing traffic.

These systems also may incorporate other pilot-interpreted information such as area navigation and aircraft-attitude displays.

ECONOMIC CONSIDERATIONS

Although the primary objective of the ATC System is to prevent collisions between aircraft, a corollary function is to expedite traffic flow. As an extreme example, if the ATC System permitted only one aircraft to fly within 1,000 miles of another, midair collisions could never take place; but, by the same token, air transportation would not exist. The challenge, therefore, is to apply collision-avoidance techniques to the extent needed to achieve an extremely high safety level, yet at the same time permitting expeditious flow of air traffic.

Air traffic delays account for a significant portion of the direct operating costs of any air service. This is particularly true of those which operate on a scheduled basis. For example, a flight scheduled for a duration of one hour block-to-block may spend only about 30 minutes in real flight time. The remaining 30 minutes are "built-in" to the scheduled time as a cushion for taxiing time and traffic delays on the ground and in the air. Over and above the "built-in" delay factor, additional traffic delays are imposed by ATC on a variable basis depending upon particular traffic conditions.

As a consequence, a typical air carrier in the United States flying an all-jet fleet with optimum cruising speeds of 550 to 600 mph will have an average system block-to-block speed of 250 to 350 mph. Although air-carrier aircraft speeds have increased about 66 percent since the introduction of jet airliners, scheduled block-to-block speeds have increased only about 20 percent. As air traffic delay pressures tend to increase, there is a parallel pressure to increase block-to-block scheduled times. This is done so that passengers feel they arrive "on time," not realizing that this is illusory in the sense that the flight time could have been considerably shorter if a large air traffic delay cushion had not been included in the scheduled trip time.

Since delays attributable to the ATC System generally are compiled only when they exceed 30 minutes of scheduled arrival time, it is obvious that *actual* air traffic delays far exceed these misleading data. This nonproductive flight time will cost a typical United States medium to medium-large airline something in the order of $5 to $7 million per month in increased operating costs.

Reducing air traffic delays can benefit the airspace users not only by reducing their direct operating costs; it also can increase aircraft productivity, thus requiring a lesser aircraft inventory to do a given job. Passengers and cargo shippers will benefit by reduced fares and tariffs, and savings in travel time.

In summary, the Air Traffic Control System must be so designed as to *prevent collisions between aircraft* and, at the same time, *expedite traffic flow to minimize economic penalties resulting from the application of collision-avoidance techniques.*

CHAPTER XII

ONLY THE BEGINNING

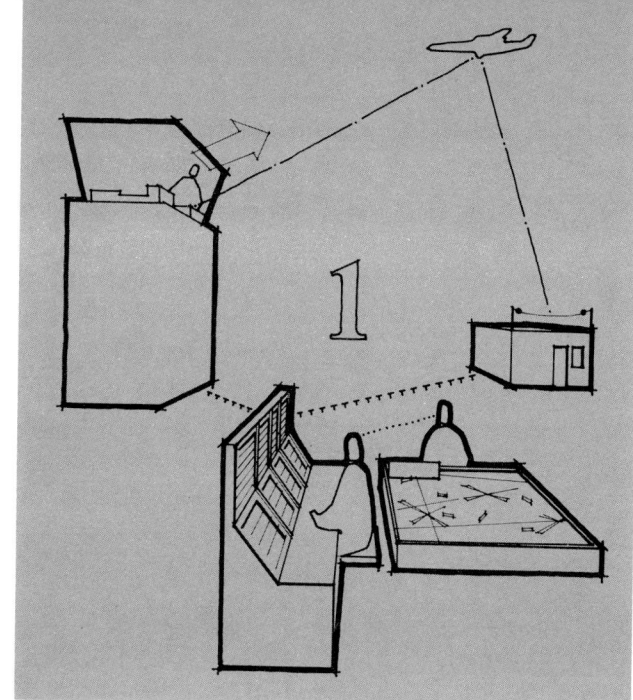

167 First-generation ATC *System was manually operated with indirect communication between controller/pilot.*

Organized means to control air traffic for collision-prevention purposes commenced a scant quarter of a century after the airplane was first flown successfully. During the years that followed, air traffic control concepts, procedures, equipment, rules, and regulations evolved to an advanced stage in the United States and internationally. A new profession had emerged—the air traffic controller. Air traffic volume of all classes had increased to proportions hardly dreamed of when the first controllers started out.

A third of a century after its initiation in the United States, the control of air traffic had become one of the most significant factors affecting air transportation development. New engineering and improvement plans were pursued. International conferences were held periodically to solve air traffic control and related problems. More and more money was being expended in operating the Air Traffic Control System. Dedicated people had devoted much time and effort to ATC System progress.

As the 1970 decade dawned, however, it was obvious that ATC facilities and system capacity had not kept up with air transportation demand in general, and particularly in areas of high-density traffic. Traffic flow was being slowed down and aircraft movements were being restricted so that air transportation demand would not exceed the ATC System capacity. Studies and analytical programs were urgently undertaken with the objective of laying the foundation for future improvements to the end that system capacity *could* keep up with system demand.

Realization slowly dawned that it was the system that was crowded—not the **Sky.**

But this is **only the beginning.** *Implementation* of worthwhile ideas must follow. The last third of the twentieth century will pose new challenges and require new solutions to problems. Innovative steps will need to be taken in order to utilize most effectively two assets which cannot be manufactured: airspace and time.

Old concepts and philosophies for the control of air traffic will need to be reviewed and reevaluated. The ATC system must be able to accommodate the constantly increasing needs of all classes of airspace users—not only with an acceptable level of performance and safety in all traffic densities, but also in an evolutionary and cost effective manner.

EVOLUTION

Air Traffic Control System "generations"—past, present and future—provide useful reference frames for systems analysis. Although applicable directly to the United States, the general concept of these generations applies throughout the world. Broadly speaking, generation definitions are:

First: mid-1930s to mid-1950s
Second: late 1940s to early 1970s
Third: early 1960s to early 1980s
Upgraded third: early 1970s to end of century
Fourth: 1980s into next century.

Historically, ATC System improvements have been evolutionary and each generation has overlapped another by some ten years. Each generation in the past has had a life span of about twenty years or so, and it is logical to assume that this general life span will apply to present and future generations.

FIRST AND SECOND GENERATIONS

The first generation is considered to have been born about 1935 with the initial efforts of several United States airlines to establish an organized "airway" traffic control system, subsequently to be taken over by the federal government in 1936. This generation was a pioneering generation, as all air traffic control procedures, techniques and equipment, rudimentary as they were, had to be developed without precedent or previous experience. Air traffic volume, relatively low during the early part of this generation, gradually increased in magnitude during its latter years. The system was completely "manually operated." Indirect communications were used between controllers and pilots, and navigation was accomplished over a somewhat limited airway system using low-frequency, four-course radio-range facilities. Another facet of the pioneering nature of this generation was the inauguration during the mid-1940s of international standards and procedures for the control of air traffic. The generation expanded from a limited "airway" area of jurisdiction to a complete air traffic control concept in-

168 Workhorse aircraft of the first generation system marked the maximum speed of air traffic during that period—about 160 knots.

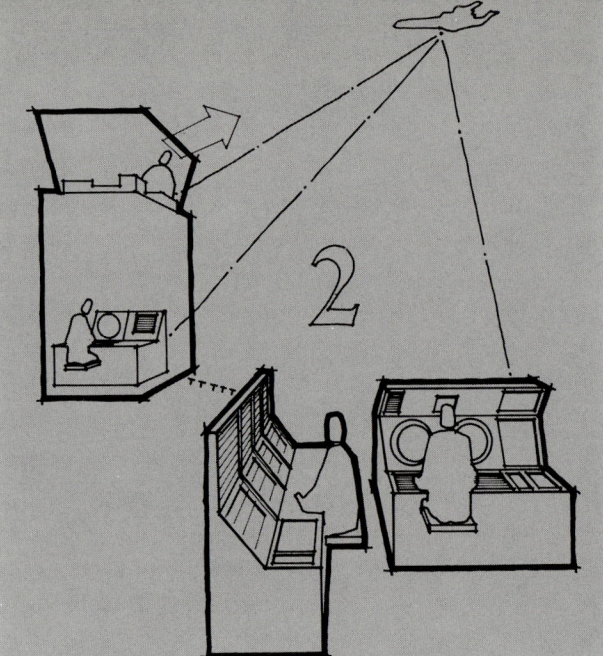

169 The advent of radar and direct controller/pilot communication links, along with improved navigation facilities, identified the advent of the second generation.

170 Aircraft during the second generation moved into the four-engine category, advancing speeds up to around 250 knots. Airborne radar transponder equipment for ATC purposes also was introduced.

The future Air Traffic Control System will make use of satellites for aircraft navigation, automatic communications with the ground facilities, and for air traffic surveillance. More capability to avoid collisions directly will be given the pilot by means of cockpit traffic situation and area navigation displays. Satellites will make possible complete global coverage

cluding en route and airport control in a consolidated system.

The advent of radar surveillance of aircraft movements in the early 1950s marked the beginning of the transition from the first to the second generation Air Traffic Control System. Included in this transition, as additional important elements, were the introduction of direct controller/pilot communications by both center and tower facilities. Modernization of aural to visual navigation capability, including the development of distance measurement equipment (DME) and instrument landing systems (ILS) was involved. There was also replacement of low, medium, and high frequencies with VHF/UHF ground and airborne communication and navigation equipment for both civil and military aviation.

The second generation carried on with and expanded the developments that had marked the beginning of the end of the first generation. Included was the consolidation of navigation concepts into a "common system" air-navigation program applicable to both civil and military aircraft—the VORTAC system. Radar became the basic tool for the control of air traffic during this generation. A primary radar system was installed, progressively composed of two basic types: airport surveillance radar (ASR) and air-route surveillance radar (ARSR). With increasing traffic, it became apparent that the controllers needed better target definition than that provided by primary radar, as well as ability to be able to identify an aircraft under surveillance without the need for the pilot to execute identifying maneuvers with the attendant communication workload.

To achieve these objectives, an ATC radar beacon system (ATCRBS) was initiated at the end of the 1950s, and implementation of both ground and airborne equipment got underway on a widespread basis during the early 1960s. The ATCRBS—also referred to as secondary surveillance radar (SSR)—provided the basis for the introduction of automation in the ATC System. Up to that time, the system continued to be manually operated just as in the first generation. With the increased speeds resulting from the introduction of jet aircraft, and the constantly increasing traffic volume, it became necessary to define specific segments of the airspace, called "sectors," to avoid exceeding controller capacity.

With increasing air traffic capacity, there also was an increasing volume of bookkeeping functions and data-processing requirements. By the end of the 1950s, an extensive network of remote communication air-ground stations (RCAG's) had been installed to provide direct controller/pilot communications with all IFR traffic, and the voice communication workload was becoming more and more of a problem. The end of the second generation system was marked by the introduction of data processing concepts.

THIRD GENERATION

The third generation system is identified basically in relation to the introduction of automation in the ground-control system. It applies on the concept of substituting mechanical processes for certain actions previously performed manually by the controller work force.

NAS Stage A

The third generation's program for the introduction of automation in the air route traffic control centers is referred to as "NAS Stage A" in the United States. The first phase of this program involves flight-data processing (FDP) which rids the center controller of certain bookkeeping and clerical tasks by printing out automatically flight progress strips derived from flight plans and aircraft position reports fed into the computer either by the controller or by external sources. This program also includes computer storage capability of flight data, flight-data updating, and automation distribution of flight data within a center's control sectors as well as between centers.

Phase two of the third generation's en route automation program consists of radar-data processing (RDP) which provides automatic aircraft tracking and computer-generated alphanumeric displays using digitized radar data. Also included in this phase is a weather subsystem display which provides the controller with the location of severe weather areas and makes possible improved weather advisory service to the pilot.

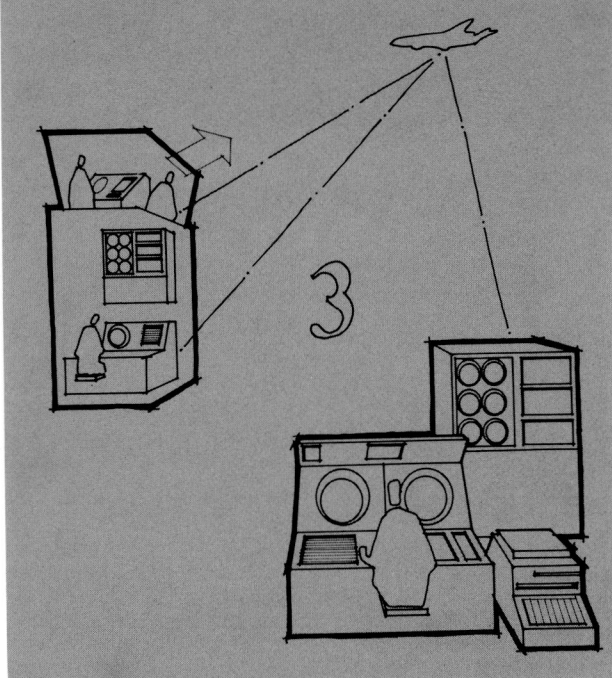

171 *The third generation initiated the era of automation in the ATC System.*

172 *Marking the third generation was the inauguration of jet propelled aircraft with speeds moving into the 500 knots category.*

The third phase in the en route automation program provides an advanced capability for the centers which will be focused on improved communication between pilot and controller, between controllers, and between the controller and the computer. Also included is provision of an accurate surveillance base for a computer-generated conflict-prediction function and a computer-generated conflict-resolution function.

The NAS Stage A automation concept is being implemented progressively during the 1970s in the United States and elsewhere, with completions projected in the 1980s.

Automated Radar Terminal System

The third generation system also ushered in a program of automation in control towers for terminal area traffic control (ARTS III) as well as at the smaller airports (ARTS II). The ARTS program provides transponder (ATCRBS) tracking and decoding capability with alphanumeric readouts showing identity, ground speed, and altitude for the suitably equipped aircraft flying in the terminal area. Additional phases of the ARTS program include primary radar tracking, digital displays, weather contours, multiple radar processing, metering, and spacing. The ARTS program also is targeted for completion by the end of the 1970s to 1980s.

Need for Upgrading

Although the third generation's system inaugurated a comprehensive automation program for the control of air traffic, by the late 1960s it became apparent that future planning of the Air Traffic Control System needed a thorough review and that perhaps the third generation system as contemplated would not be adequate to meet the traffic demands forecast for the 1970s and 1980s. As a consequence, the United States government formed an Air Traffic Control Advisory Committee (ATCAC) for the purpose of recommending an air traffic control system for the 1980s and beyond. This committee completed its studies by the end of 1969 and proposed a program for "upgrading" the third generation system in the light of forecast air traffic developments. The study concluded that "the demand for all categories of aviation will maintain its high growth rate unless further constrained by an inadequate Air Traffic Control System. The various national entities of aviation activity are predicted to at least double by 1980 [with respect to 1970] and to double again by 1995."

UPGRADED THIRD GENERATION

Probably the first significant step to mark the transition from the third generation as originally conceived to the upgraded third generation was the advent of area navigation (RNAV). This marked a shift in the previous ATC System philosophy in that by use of RNAV the pilot can assume a greater degree of direct responsibility for navigating his aircraft, and the controller's workload is

173 *Forecast air traffic increases indicate a potential Air Traffic Control System capacity gap in relation to traffic demand during the last half of the 1970s and first part of the 1980s until an "advanced" or fourth generation ATC System can be implemented. An "upgraded" third generation system will fill this gap.*

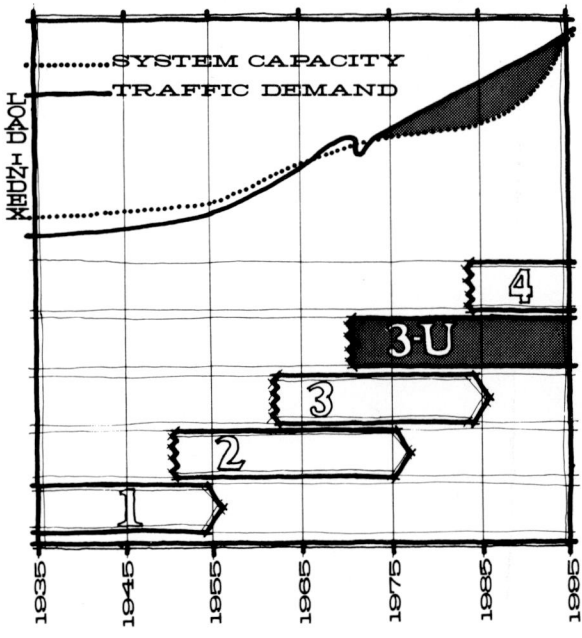

correspondingly reduced. In other words, RNAV permits distributing certain air traffic management functions from the controller to the pilot, thus helping to "unload"

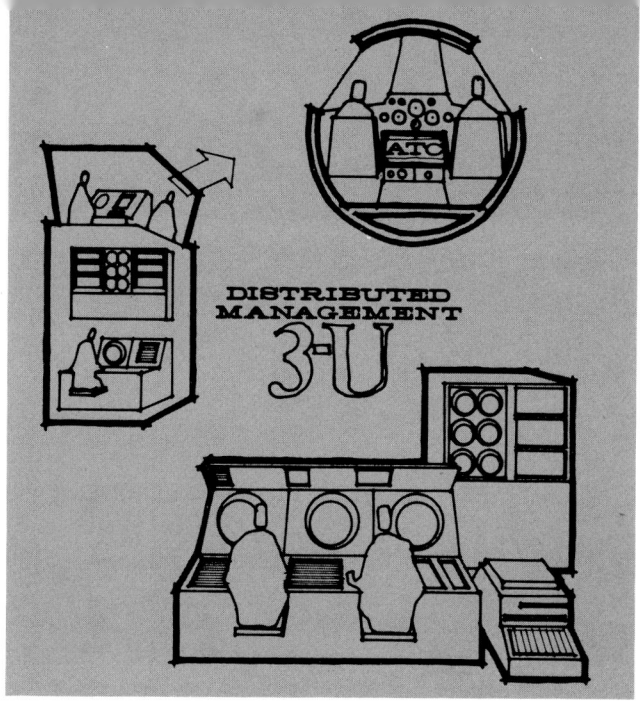

174 The upgraded third generation system as it emerges during the 1970 decade will provide more distribution of workload between controller and pilot, as well as advanced automation and other technological applications.

the ground system. One of the fundamental philosophical questions being resolved in the development of the upgraded third generation system during the mid to latter 1970s is the extent to which traffic-management functions may be distributed most advantageously between the controller and the pilot.

Comparison with Third Generation

The following table provides a comparison between the United States' basic third generation and the proposed upgraded third generation in terms of principal air traffic management functions.

AIRCRAFT SEPARATION *Sequencing, Spacing*

Third generation	Upgraded third generation
	Improved beacon/secondary radar system
	Upgraded ATCRBS
	Discrete address beacon system (DABS)
	Beacon/secondary radar backup
	Primary radar
Primary radar	Procedural
	See and avoid
Radar beacon (ATCRBS)	Manual radar vectoring
	ARTS II and III data acquisition (terminal)
See and avoid	NAS Stage A data acquisition (en route)
	Automated data processing
Procedural	Automated metering and spacing
	Automated conflict prediction and resolution
Manual radar vectoring	Expanded positive control areas (PCA)
	Airspace organization
	VFR only
	IFR only
	Performance segregation
	Mixed

FLOW CONTROL

Third generation	Upgraded third generation
Manual computations	Automated central traffic flow management
	Gross prediction of high-density terminal demand and capacity
Manual central-flow-control management	Flow-rate control
	Automated local (center area) flow control
	Short-term demand and capacity prediction
	Flow-rate control

NAVIGATION

Third generation	Upgraded third generation
	Airborne Area Navigation
VORTAC	VORTAC
	VOR/DME
Doppler VOR	PVOR/DME
	Doppler
TVOR	Inertial
	Hyperbolic
Manual radar vectoring	Pre-organized RNAV routes (2-D and 3-D)
	Manual radar-vectoring backup

175 Typical of the new aircraft identified with the upgraded third generation are the advanced jumbo jets, advanced STOL's and helicopters, and the SST.

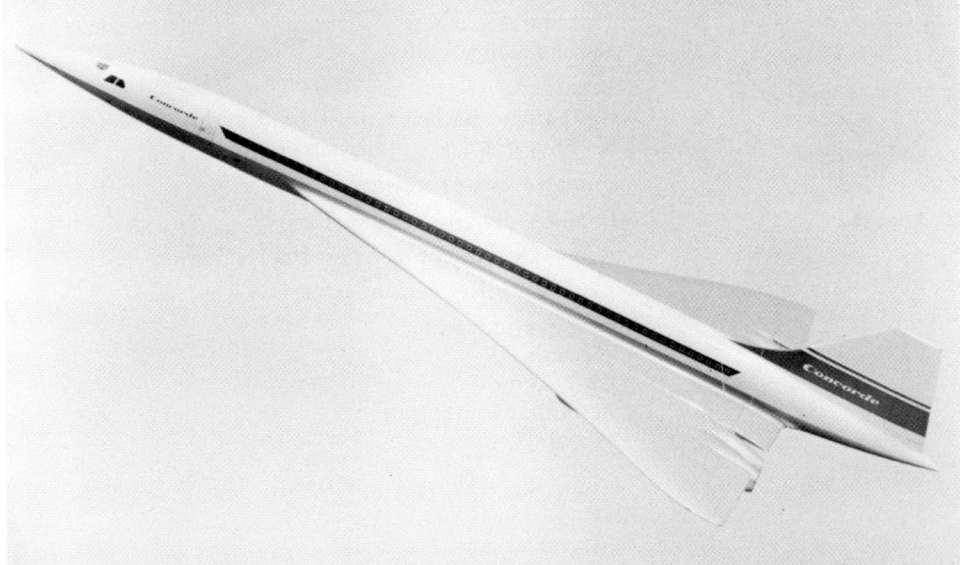

APPROACH GUIDANCE

Third generation	Upgraded third generation
Standard ILS	Standard ILS
	Microwave ILS (standard and scanning,
Manual radar vectoring	3-D/4-D RNAV
	Lighting
Lighting	Manual radar vectoring

COMMUNICATIONS

Third generation	Upgraded third generation
	Data link
	Position
	Speed
Voice communication	Heading
	Altitude
Teletype	Voice communication backup
	Discrete frequency
Manual FSS advisory	Satellite
	Oceanic
Manual preflight briefing	Contiguous
	High-speed teletype
	Automated FSS advisory
	Automated mass preflight briefing
	Intermittent positive control (IPC)

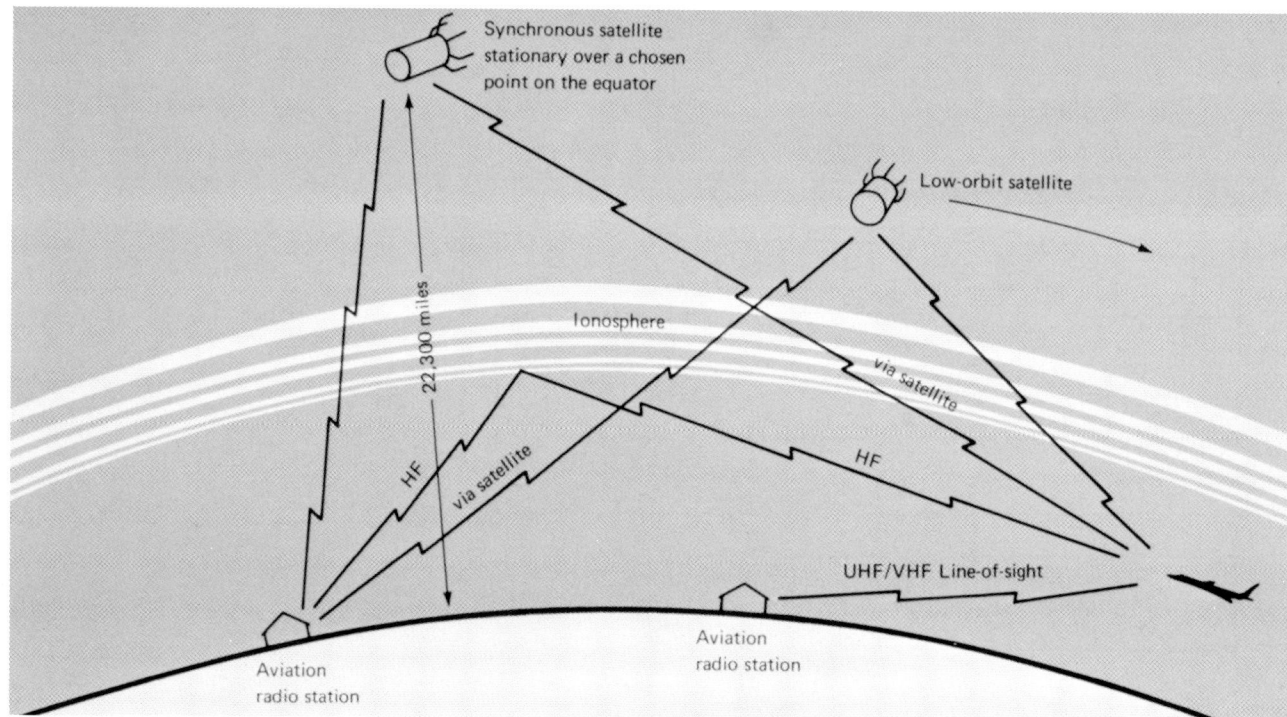
176 Application of satellites for aeronautical communications purposes in upgraded third generation system.

From the foregoing comparison, it will be seen that by the end of the 1970s or early 1980s, the third generation will have merged into the upgraded third generation. No doubt during this transition period, some modifications to the upgraded third generation concept will be made in the light of evolutionary experience and changing airspace users' requirements.

Analysis

The following is an analysis of the principal elements comprising the upgraded third generation system as visualized by 1980.

• Traffic management will remain essentially as a man-intensive, ground-centralized function, although some navigation capabilities will be delegated to the pilot through the application of area navigation concepts. Basically, however, management of air traffic will continue to rely on radar surveillance, with the controller in most situations of medium- or high-density traffic continuing to control aircraft individually by radar vectoring. More aircraft will be brought under the jurisdiction of the ground traffic-management system by means of expanding positive control areas. Airspace structuring will tend to segregate different classes and categories of air traffic.

• Ground automation facilities for data processing and alphanumeric radar displays will be emphasized. Expanded ground automation will apply to centers, towers, flow control, and flight service stations including automated mass preflight briefing of pilots. Other expanded automation functions will include computer capability to provide controller assistance for metering and spacing aircraft as well as conflict prediction and resolution.

• Radar will continue as the one and only source for automatic or semiautomatic flight-data acquisition. The ATC radar beacon system (ATCRBS) will be improved, but is scheduled to be replaced by a discrete address beacon system (DABS). Primary radar will be continued as a backup for aircraft not equipped with the ATCRBS or DABS.

• DABS is the essential heart of the entire upgraded planning, and is programmed for operational use by the late 1970s. DABS will be provided with data-link capability to receive digital transmissions from the ground Air Traffic Control System on a discrete address basis. With DABS, intermittent positive control (IPC) is planned which would permit the ground system to transmit data-link instructions to a particular aircraft only as required for collision-avoidance purposes, i.e., altitude, track, or speed changes.

• Navigation ground facilities will continue basically as in the third generation, except for some possible improvements in accuracy. Airborne self-contained systems will be brought into greater use. Airborne area navigation will become the basic system, with area navigation routes and airspace structuring progressively replacing the conventional airways system by the early 1980s. Microwave instrument-landing systems, both standard and scanning beam, will be introduced on an increasing basis to replace or supplement the conventional ILS facilities.

• Communications will shift more to data-link applications. Voice communications will continue as a backup, with discrete frequencies being used for specific communications purposes. Satellites will be introduced for oceanic communications and perhaps to some extent over the continental United States. Weather information and flight-plan distribution will be expedited via high speed teletype circuits.

Problem Areas

Although the upgraded third generation contributes significantly to improving the basic third generation, certain problem areas will need to be considered during its evolutionary development. These may be resolved as part of the upgraded third generation system, or perhaps as part of the transition into the fourth generation system. Included in these problem areas are the following general subjects.

• Radar Deficiencies. Basic radar deficiencies were outlined in Chapter 6. These deficiencies bring into question whether the DABS and IPC concepts can function effectively in high-density traffic areas, or what improvements to radar surveillance will be needed to permit them to function effectively. Other points of concern include deficien-

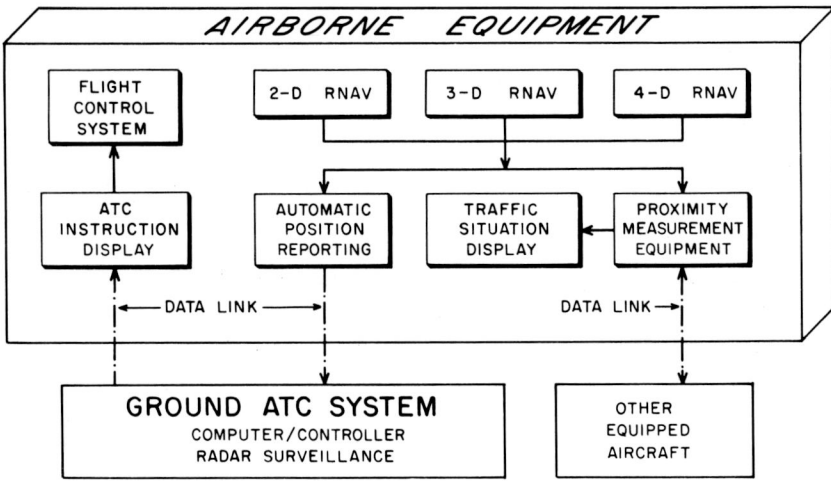

177 High speed automatic communications can link airborne navigation equipment with ground ATC computers to provide radar surveillance backup/redundancy, as well as to provide ground/air data link inputs to suitable cockpit displays.

cies in radar to provide coverage of all airspace down to the surface, and lack of system redundancy in the event of significant radar failure.

• **Centralized Traffic Management.** Continued reliance on essentially a 100 percent ground-centralized, traffic-management system will require a constantly increased controller work force and more sectors with increased sector-coordination problems. Both of these factors will lead to a counter-productive situation in terms of Air Traffic Control System capacity and operating costs. This counter-productive situation will be compounded further by the planned increases in positive controlled airspace. While automation concepts materially reduce controller bookkeeping and routine workload, by themselves they do not significantly increase airspace and airport capacity.

• **Pilot Participation.** Aside from the introduction of area navigation applications, the upgraded third generation system does not contemplate bringing the pilot significantly into the air traffic management decision-making loop. In looking to the future design of the ATC System, perhaps the single most decisive conceptual question to be answered is the extent to which traffic management in the system should be distributed between the pilot and the controller. Should the pilot be able to play an active part in the traffic management process, or just a passive role? Should a new generation of cockpit instrumentation, such as traffic situation displays, be de-

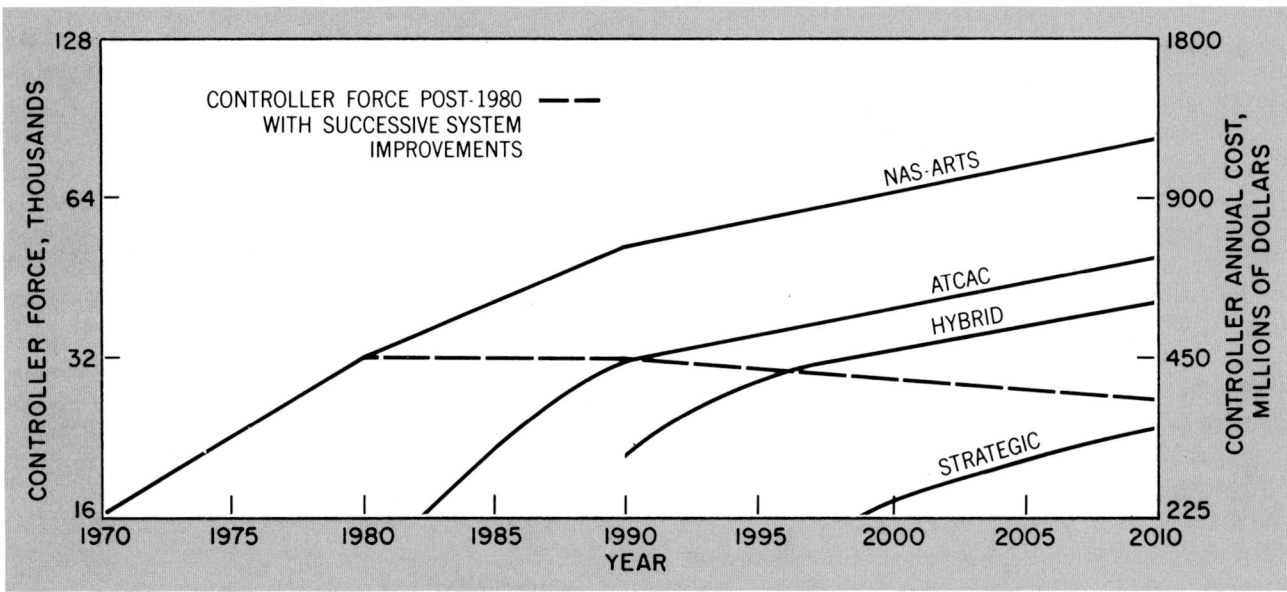

178 Controller work force requirements related to different ATC System concepts.

veloped to give the pilot redundant capabilities for the purpose of traffic separation assurance and spacing?

• **Improved Navigation.** Accurate airborne area navigation capability, with no line-of-sight restrictions, is needed in order to open up *all* airports for IFR capability, with the resultant positive impact on all classes of air transportation including stimulating economic development of rural areas, facilitating IFR use of all runways at any airport, and expanding city center and local/short-haul air service. The reliance on VOR/DME (VORTAC) as the basic navigation system in the upgraded third generation system planning does not appear to offer the accuracies and flexibilities needed to support the efficient admixture of CTOL, STOL, and VTOL type aircraft. Looking toward the future, navigation systems are needed which

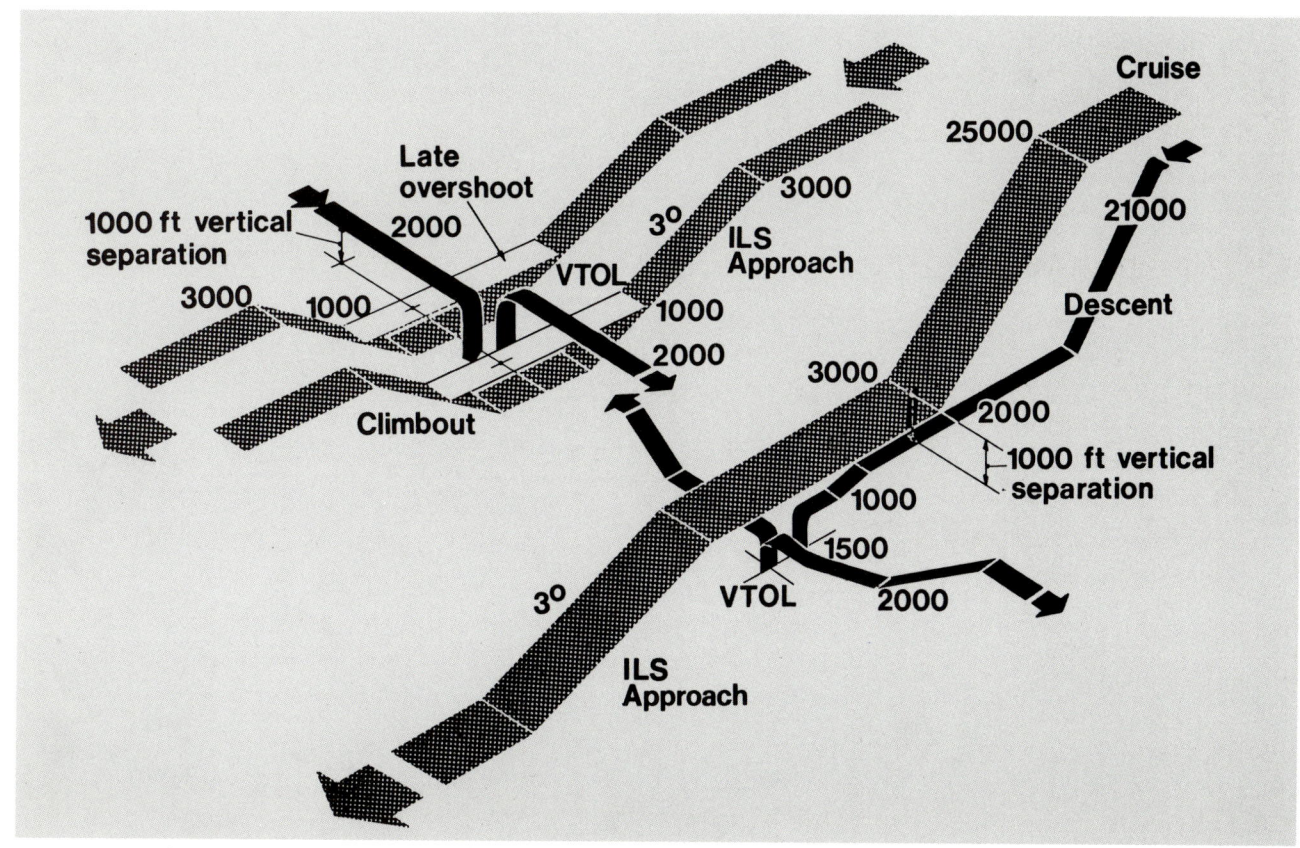

179 Extremely accurate 3-D and 4-D navigation systems can make possible V/STOL operations completely independent of CTOL traffic in high-density environments.

can provide accuracies in three dimensions to within a few feet, on an area basis and down to the surface. Should an extremely accurate ground/air area navigation system be developed, then extensive installations of ground-instrument landing systems to serve each runway at every airport desiring "all-weather" service could be significantly minimized.

• Airport/Airspace Productivity. Probably the fundamental question to be answered in evaluating planning for the upgraded third generation ATC system is the extent to which it will contribute to collision avoidance and at the same time be cost effective. Related to this question is the need to provide greater airport productivity and terminal area efficiency.

• The V/STOL. This type of air vehicle offers a major opportunity to move forward in using airports and airspace more productively. The ATC System challenge is to permit these air vehicles to utilize most effectively **The Uncrowded Sky** as a transportation medium.

The Next Steps

Recognizing that the upgraded third generation system was not in itself a complete solution to the Air Traffic Control System requirements anticipated for the 1980s and 1990s, studies were inaugurated in the early 1970s looking toward the development of a fourth generation or "advanced air traffic management system." As in the past, the transition from the upgraded third generation to the fourth generation will be evolutionary and require perhaps a ten-year transition period where one generation will merge into the next.

FOURTH GENERATION

In the fourth generation Air Traffic Control System, certain goals need to be identified. These goals can be defined quite well by projecting the problem areas of the upgraded third generation system and by postulating how the fourth generation can best solve these problems, as well as meeting the additional challenges which may be forecast for the latter part of the century. Fundamental to the approach is how to minimize transition difficulties, at the same time achieving goals in the most effective manner.

Controller/Pilot Responsibility

The number one human-factor problem in moving forward into future Air Traffic Control System design is that of controller/pilot responsibility. Behavior of the *man* in the system must be given high priority. Related directly to this question is the problem of man/machine interface.

For example, a system configuration which requires that the controller acknowledge the existence of each and every aircraft within his area of concern certainly implies a high level of responsibility, particularly as traffic densities increase to those forecast in the late 1970s and beyond. By the same token, increasing traffic density fo-

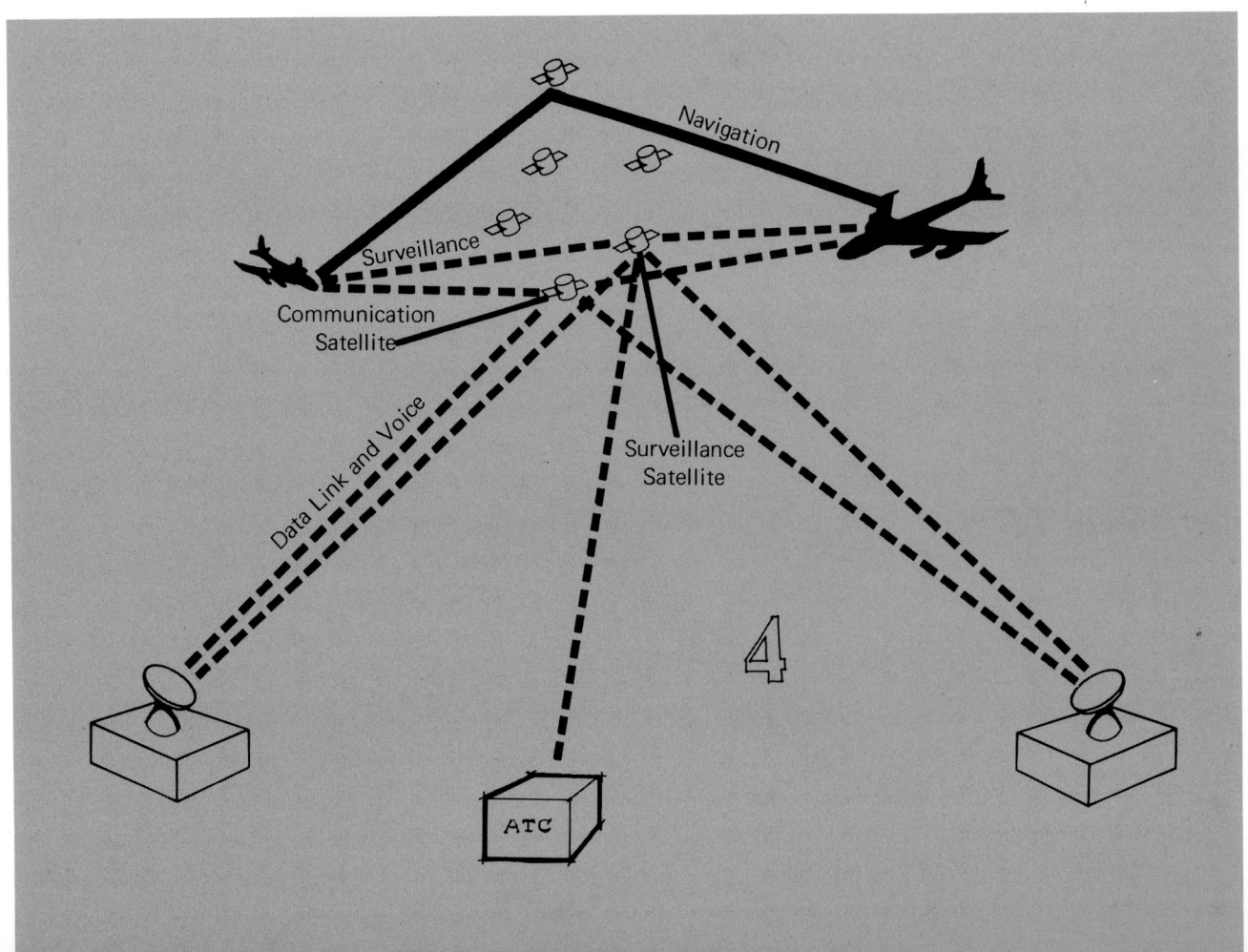

180 Fourth generation (advanced air traffic management) system will use satellites in combination with ground stations.

181 Super-jumbo CTOL jets will become part of the ATC System during the fourth generation.

182 Subsonic VTOL's with long-range capability will usher in the fourth generation, combining vertical and takeoff lift with high-speed performance.

cuses more and more attention on the need for the pilot to have more independent capability of determining the destiny of his craft in the air traffic environment.

Human factor considerations, such as those relating to controller and pilot responsibility, may comprise the basis for the most difficult problems to solve in the development of an advanced air traffic management system. These same considerations, of course, apply to the evolutionary development of the upgraded third generation Air Traffic Control System during the 1970s and into the 1980s.

Distributed Management Concept

In its report on the long range needs of aviation submitted to the president and Congress in the first part of 1973, the United States Aviation Advisory Commission—created by the Airport and Airway Development and Revenue Act of 1970—defined distributed "air traffic management" as being "a system in which the ground controller is responsible for overall traffic flow, routing and monitoring, while the individual pilot is responsible for navigation, stationkeeping and collision avoidance in conformance with an approved flight plan."

The commission pointed out that such a system requires that the precise position and identification of the aircraft involved and of all other aircraft in its area of concern be displayed electronically in the cockpit, as well as to the ground controller, by means of a device which may be called "an air traffic situation display (ATSD)."

Candidate Concepts

In postulating a fourth generation or advanced air-traffic-management system, there are three basic concepts which may be considered.

• *A distributed management system* in which a maximum of control functions is allocated to the aircraft.

• *A centralized management system* in which a maximum of control functions is allocated to the ground, and guidance is accomplished with heading and speed commands to the airplane. (This concept is referred to as "tactical" control.)

• *A centralized management system* in which control functions are divided between the aircraft and the ground, and guidance is accomplished by providing long-term, four-dimensional, flight-path clearances to the airplane. (This is referred to as "strategic" control.)

Combinations of these basic three concepts will be involved in shaping the fourth generation system. In any event, looking toward the continuing development of computer technology, greater use of on-board data processing offers much promise for the future. In fact, the real impact on the concept of future air traffic management design no doubt will rest to a significant extent on the degree to which inexpensive, on-board, data-processing computers are introduced into the Air Traffic Control System.

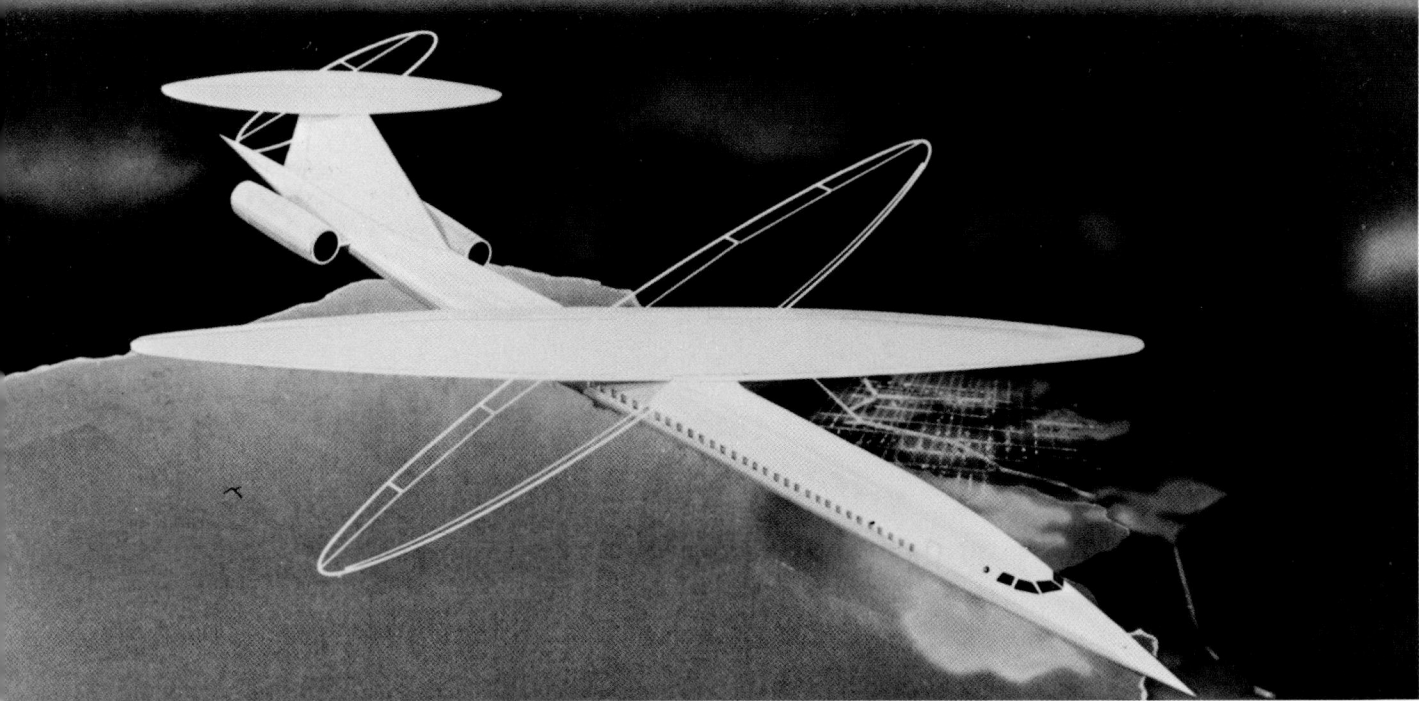

183 *New aerodynamic designing could produce a quiet, nonsonic boom* SST *as one of the new air vehicles to be introduced in the fourth generation* ATC *System.*

Supporting Concepts

Whatever the philosophical approach may be in designing the fourth generation system, a fundamental factor is the need for a new highly accurate navigation system that will be part of an "integrated communication, navigation, and identification (ICNI)" concept. Time/frequency (T/F) technology appears to offer virtually unlimited capability to meet this objective. The same technology can be used to supply air traffic situation displays and other airborne tools, such as CAS and perhaps PWI devices, to assist the pilot/controller team. T/F facilities may be located on the ground, or they may use satellites to provide global coverage. Other candidate navigation systems may include differential Omega and LORAN, involving subfacilities situated in specific areas to compensate for local variations.

Final Challenge

Probably the final challenge in designing the fourth generation/advanced air traffic management system will be the success with which all of the system elements are taken into consideration as a whole: the users; the pilots and the controllers; the ground and airborne facilities and systems; rules, regulations, and procedures; the aircraft; the airports; and the airspace.

Some of the "old" controllers will be around for the fourth generation and the fiftieth anniversary of the founding of air traffic control. None will be here for the one-hundredth anniversary. But the mantle of enthusiasm, inventiveness, and pioneering spirit is passed on to all future controllers.

ACKNOWLEDGMENTS

The outstanding cooperation of the Federal Aviation Administration in the preparation of this book is acknowledged with sincere appreciation. The author is especially grateful to the FAA's Air Traffic Service for technical inputs and manuscript review, to the Office of Aviation Policy and Plans for FAA policy-planning illustrative material, and to the Systems Research and Development Service for engineering information. Also to various other divisions of the FAA for reference material, including the Quiet Short-haul Air Transportation System Office, Flight Standards Service, Office of Systems Engineering Management, and the National Aviation Facilities Experimental Center.

A well-merited acknowledgment goes to my executive secretary and wife, Martha Gilbert, for typing a mountain of manuscript drafts as well as classifying tons of research material; also to Gordon Gilbert, architect, for many of the special illustrations contained in the book.

Other contributors who supplied valuable information include:

Aeronautical Radio, Inc.
Aerospace Industries Association
Aerospatiale
Airbus Industrie (Europe)
Aircraft Owners & Pilots Association
Airline Dispatchers Association
Air Line Pilots Association
Airport Operators Council International

Air Traffic Control Association
Air Transport Association of America
American Air Lines, Inc.
American Institute of Aeronautics & Astronautics
Association of Local Transport Airlines
Autonetics-North American Rockwell
Aviation Advisory Commission

Beech Aircraft Corporation
Bell Helicopter Company, Division of Textron
The Bendix Corporation
The Boeing Company
Curtiss C. Bogart
Bolkow GmbH (Germany)
British Aircraft Corporation, Ltd.
British Air Line Pilots Association
Britten-Norman, Ltd. (England)
Brandt Driftoff Runway Co.
Butler National Corporation

CAE Industries, Ltd. (Canada)
Canadian Aeronautics & Space Institute
Canadian Air Traffic Control Association
Canadian Department of Industry
Canadian Department of Transport
Canadian Marconi Co.
Canadair, Ltd.
Cessna Aircraft Corp.
Champlain Technology, Inc.

Civil Aeronautics Board
Homer F. Cole (Historical Records)
Collins Radio Co.
Communications Satellite Corp.
Compagnia Nazionale Aeronautica (Italy)
Cornell Aeronautical Laboratory
Country Air Parks, Ltd. (England)

Decca Navigator Co., Ltd.
The de Havilland Aircraft of Canada, Ltd.
DELCO Electronics
Der Bundesminister fur Verkehr (Germany)
Dornier, AG (Germany)

Eastern Air Lines, Inc.
Eurocontrol

Fairchild Industries
Ferranti-Packard Electric, Ltd. (Canada)
Ferranti Digital System Dept. (England)
Ferranti, Ltd. (Scotland)
FIAT-Divisione Aviazione (Italy)
Flight Safety Foundation
Florida State Department of Transportation
Fokker-VFW, N.V. (The Netherlands)
French Ministère de l'Equipement, Secrétariat General de l'Aviation Civil
Ft. Worth (Texas) City Planning Department

Garrett Manufacturing Ltd. (Canada)

Gates-Learjet Corp.
General Aviation Manufacturers Association
General Electric Company
Global Navigation, Inc.
Grumman Corporation
Guild of ATC Officers of Great Britain

Hawker Siddeley Aviation, Ltd.
Helicopter Association of America
N V Hollandse Signaalapparaten

IBM Corporation
Insurance Company of North America
International Aircraft Owners & Pilots Association
International Air Transport Association
International Civil Aviation Organization
International Federation of Air Traffic Controllers Associations
Institute of Electrical & Electronics Engineers
International Federation of Air Line Pilots Association
Italy - **Ministero dei Trasporti e della Aviazione Civile**

Jeppesen & Co.
Jane's All the World's Aircraft

King Radio Corporation
KLM-Dutch Airlines

Lear-Siegler, Inc.
·Life Magazine

Litchford Systems
Litton Industries, Inc.
Lockheed Aircraft Corp.
Lockheed California Co.
Lockheed Georgia Co.
Los Angeles (California) Department of Airports
LTV, Inc.
Lufthansa-German Airlines

The Marconi Co., Ltd. (England)
Massachusetts Institute of Technology, Dept. of Aeronautics & Astronautics
McDonnell-Douglas Corporation
Miami (Florida) Chamber of Commerce-Aviation Division
Minneapolis-Honeywell Corp.
The Mitre Corporation

NARCO Avionics
National Air Transportation Conferences
National Association of State Aviation Officials
National Aviation Trades Association
National Business Aircraft Association
National Transportation Safety Board
NATO Committee for European Airspace Cooperation (CEAC)
National Aeronautics & Space Administration (NASA)
NASA-Ames Research Center
NASA-Langley Research Center

New York Airways
Norden Division-United Aircraft
New York State Department of Conservation
North American Rockwell Corp.
Northrup Corporation

Ocean Measurements, Inc.

Pan American World Airways
Piper Aircraft Corporation
Professional Air Traffic Controllers Organization

Radio Technical Commission for Aeronautics
Raymond Engineering, Inc.
Raytheon Canada, Ltd.
Ransome Airlines
RCA
Rolls Royce, Ltd.
Royal Aeronautical Society of Great Britain

SABENA-Belgian Airlines
Selenia (Italy)
Short Brothers & Harland, Ltd. (Northern Ireland)
Sierra Research Corp.
Singer-Kearfott Division
Sikorsky Aircraft Division
Sperry Rand Corp.
Standard Radio & Telefon AB (Sweden)
Standard Telephones & Cables, Ltd. (England)

Stanford Research Laboratory
Sud-Aviation (France)
Syracuse University Research Corp.

Teledyne CAE
Telefunken (Germany)
Texas Instruments, Inc.
Trans World Airlines, Inc.
TRW, Inc.

United Aircraft Corp.
United Air Lines, Inc.
United Kingdom Ministry of Defence-National ATC
  Services
United Kingdom Board of Trade-Civil Aviation Dept.
United States Government
  Department of Commerce
  Department of Defense, U.S. Air Force, U.S. Army,
    U.S. Navy
  Department of Transportation, U.S. Coast Guard,
    Federal Aviation Administration, Transportation
    Systems Center

VFW (Vereinigte Flugtechnische Werke) (Germany)

James Watson
Wilcox Electric Co.
World Meteorological Organization

Stanford Research Laboratory
Sud-Aviation (France)
Syracuse University Research Corp.

Teledyne CAE
Telefunken (Germany)
Texas Instruments, Inc.
Trans World Airlines, Inc.
TRW, Inc.

United Aircraft Corp.
United Air Lines, Inc.
United Kingdom Ministry of Defence-National ATC
  Services
United Kingdom Board of Trade-Civil Aviation Dept.
United States Government
  Department of Commerce
  Department of Defense, U.S. Air Force, U.S. Army,
    U.S. Navy
  Department of Transportation, U.S. Coast Guard,
    Federal Aviation Administration, Transportation
    Systems Center

VFW (Vereinigte Flugtechnische Werke) (Germany)

James Watson
Wilcox Electric Co.
World Meteorological Organization